Ethnobiology

Ethnobiology is the study of the dynamic relationship between plants, animals, people, and the environment. Academic and applied interests include ethnobotany, ethnozoology, linguistics, paleoethnobotany, zooarchaeology, ethnoecology, and many others. The field lies at a dynamic intersection between the social and biological sciences. The major contribution from the biological sciences has come from economic botany, which has a rich historical and scientific tradition. Indeed, the objectives of the colonial enterprise were as much about the quest for "green gold" –herbal medicines, spices, novel cultivars, and others—as it was for precious metals and sources of labor. The view that ethnobiology concerns mostly the discovery of new and useful biota extended into the 20th century. The social sciences have contributed to the field in both descriptive studies but also within quantitative approaches in cognitive anthropology that have led to general principles within ethnobiological classification. Ethnobiological research in recent years has focused increasingly on problem solving and hypothesis testing by means of qualitative and especially quantitative methods. It seeks to understand how culturally relevant biotas are cognitively categorized, ranked, named, and assigned meaning. It investigates the complex strategies employed by traditional societies to manage plant and animal taxa, communities, and landscapes. It explores the degree to which local ecological knowledge promotes or undermines resource conservation, and contributes to the solution of global challenges, such as community health, nutrition, and cultural heritage. It investigates the economic value and environmental sustainability to local communities of non-timber forest products, as well as the strategies through which individual ecological knowledge and practices encourage resilience to change—modernization, climate change, and many others. Most importantly, contemporary ethnobiological research is grounded in respect for all cultures, embracing the principles of prior informed consent, benefit sharing, and general mindfulness.

More information about this series at http://www.springer.com/series/11551

Anton Daughters • Ana Pitchon
Editors

Chiloé

The Ethnobiology of an Island Culture

Springer

Editors
Anton Daughters
Department of Sociology, Anthropology, and Justice Systems
Truman State University
Kirksville, MO, USA

Ana Pitchon
Department of Anthropology
California State University Dominguez Hills
Carson, CA, USA

ISSN 2365-7553 ISSN 2365-7561 (electronic)
Ethnobiology
ISBN 978-3-319-91982-9 ISBN 978-3-319-91983-6 (eBook)
https://doi.org/10.1007/978-3-319-91983-6

Library of Congress Control Number: 2018946647

Printed on acid-free paper

This Springer imprint is published by the registered company Springer International Publishing AG part of Springer Nature.
The registered company address is: Gewerbestrasse 11, 6330 Cham, Switzerland

Acknowledgments

The list of active anthropologists and geographers who have carried out ethnographic research in Chiloé is not long, and this volume represents a wonderful collaboration between one such group. We would like to thank our friends and colleagues in Chiloé who have helped us with our work over the years. These include Renato Cárdenas Álvarez, Daniela Leviñanco, Ana Rosa Uribe, Ramón Contreras, Lorna Muñoz, Irene and Hugo Mansilla, Raquel Mansilla, Armando Bahamonde, Doris Barría, Fabián Bahamonde, Pamela Urtubia, Marijke van Meurs, Miguel Jiménez, Edith Mansilla, and Carlos Aynol. Our understanding of Chiloé would be far more limited without their hospitality, solidarity, and knowledge. We also thank the editors at Springer—Eric Stannard, Rick Stepp, and Rahul Sharma in particular—who have helped shepherd this project through its various stages.

Ana Pitchon gratefully acknowledges Pedro Cardenas, Erica Antiñanco, and Alejandro Cardenas Antiñanco, who introduced her to the true spirit of Chilote cooperation. "I am forever grateful for their friendship, warmth, stories and all the mussels I could ever eat!" writes Ana.

Sarah Ebel thanks the US National Science Foundation Adaptation to Abrupt Climate Change IGERT program for their funding support of her project. She would also like to thank research assistant Maribel Bustamante Toro for her hard work and friendship.

Richard Vercoe extends his appreciation to the communities of Meulín Island and Comau Peninsula for sharing their time, patience, knowledge, and beautiful piece of the world with him. He would also like to acknowledge Dr. William Gribb at the University of Wyoming for his guidance and support of his research on the *trueque* Chilote.

Eric H. Thomas wishes to thank Drs. Rudi Colloredo-Mansfeld and Florence Babb for looking at early versions of his research and for providing valuable feedback, as well as Jorge Padilla Trujillo, DVM, for his help in the field.

Contents

Contributors

Giovanna Bacchiddu Programa de Antropología, Pontificia Universidad Católica de Chile, Santiago, Chile

Anton Daughters Department of Sociology, Anthropology, and Justice Systems, Truman State University, Kirksville, MO, USA

Sarah A. Ebel Department of Anthropology, University of Maine, Orono, ME, USA

Ana Pitchon Department of Anthropology, California State University Dominguez Hills, Carson, CA, USA

Eric H. Thomas Department of Anthropology, University of North Carolina, Chapel Hill, NC, USA

Richard A. Vercoe Department of Geography, University of Wyoming, Laramie, WY, USA

Chapter 1
Chiloé Today and Over the Centuries

Anton Daughters

Introduction

During the annual meeting of the Society for Ethnobiology in Denver in 2012, the ornithologist Amadeo Rea was asked to comment on the importance of ethnobiology. Rea spoke of the intense ecological familiarity that develops out of long-term dependence on a particular environment. "You have people who have lived in a place for centuries, if not millennia, and who have adapted so well. That intimate universal community knowledge of natural history and environment is such a beautiful [thing]… a work of art if you are looking at the whole cultural ecosystem. [It is] efficient and functional and sustainable…. And it is a vanishing resource" (Gosford 2013).

There are few places in the world today where Rea's words are more fitting than the islands that make up southern Chile's Archipelago of Chiloé. People have occupied Chiloé for millennia, drawing from the ocean, beaches, and cold-weather rainforests for survival. They have developed an intimate understanding of that natural landscape through subsistence farming, fishing, and aquaculture. And they have retained much of that knowledge even as parts of the region become increasingly industrialized. Nature looms large in their livelihood practices, cultural traditions, myths, and ethos.

This book explores the human-biota relationship in this part of Chile. It presents the work of anthropologists and geographers who have carried out ethnographic field research on the islands that make up Chiloé proper and on the mainland across from the archipelago. It is built around the following three questions: How do the rural residents of southern Chile draw from the environment to meet their everyday needs? How is the natural world reflected in their culture? And how is that

A. Daughters (✉)
Department of Sociology, Anthropology, and Justice Systems, Truman State University, Kirksville, MO, USA
e-mail: adaughters@truman.edu

A. Daughters, A. Pitchon (eds.), *Chiloé*, Ethnobiology,
https://doi.org/10.1007/978-3-319-91983-6_1

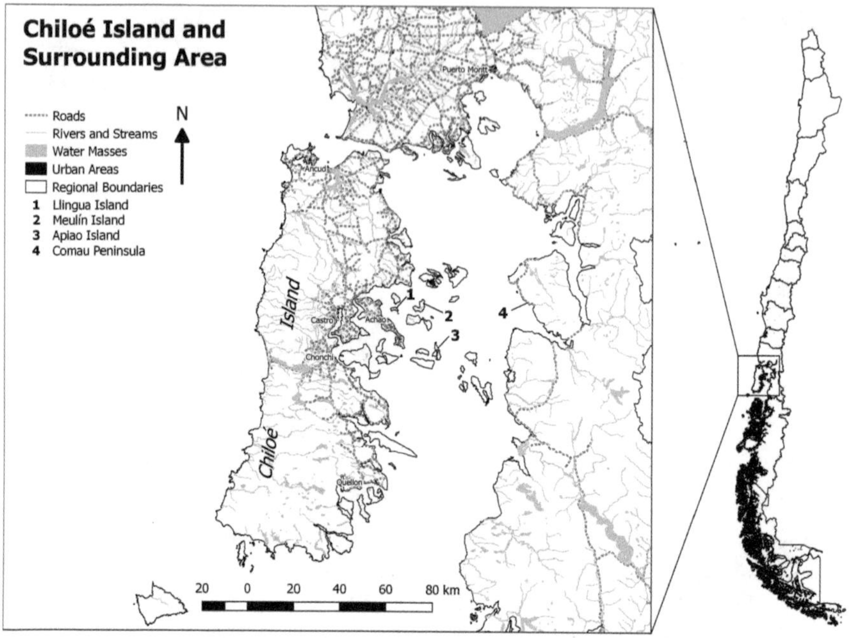

Fig. 1.1 Map of Chiloé's big island and surrounding area. (Map credit: Kurtis A. Butler and Richard A. Vercoe)

relationship changing? The chapters offer a mix of qualitative and quantitative research undertaken between 2005 and 2017. In the spirit of Gary Paul Nabhan's urging to pay attention to every generation's "cultural sense of place," they emphasize local knowledge and practice and, wherever possible, the voices of islanders (2016:144) (Fig. 1.1).

The Place

Chiloé consists of several dozen islands located approximately 700 miles south of Chile's capital city Santiago. The name of the archipelago comes from the Huilliche word *chilhue*, meaning "place of the seagulls." Its largest island—the *Isla Grande*—is approximately the size of Puerto Rico, with smaller islands tightly clustered alongside it. Along its northern edge, the Isla Grande comes within two miles of mainland Chile, separated by a narrow channel called Chacao. Along its eastern side, it is separated from the Andes Mountains by a 30-mile wide interior sea. On clear days, one can see from the hilltops the jagged snowcapped peaks of the mainland, including the volcano Corcovado that marks the northern range of Patagonia.

Undeveloped portions of Chiloé's islands are covered in dense temperate rainforests, part of the larger Valdivian forest ecozone that also covers much of southern

Chile's Patagonia. Most of the archipelago's settlements are located along the eastern (inland) sectors, where the hills are lower and the ocean less turbulent, and where large swaths of the land have been cleared for agriculture and livestock pastures. The beaches and forests in these parts of the archipelago are dotted with giant rhubarb-like plants (*Gunnera tinctoria*), known locally as *nalcas*—used for traditional meals and medicine—and a wild bush called *quiscal*,[1] used extensively for basket weaving. Fields are often covered by a very aggressive invasive bush known as *espinillo*.[2] Imported by European settlers in the nineteenth century, the bush produces yellow flowers that give Chiloé's roadsides a characteristic look in the summers. The landscape of lush pastures, dense evergreen forests,[3] and colorful wood-tiled homes often lead visitors to compare rural Chiloé to J.R.R. Tolkien's mythical land of the Shire.

Chiloé's western, Pacific-facing coast is far less populated. Buffeted by strong winds, the coastline is a mix of bluffs and rocky beaches. Cucao, the largest town on this side of the Isla Grande, or big island, is home to about 500 people. Much of the old growth temperate rainforests surrounding Cucao are protected by Chiloé National Park, a government-run preserve established in 1982. A second natural preserve, Parque Tantauco, is owned by the wealthy businessman and current president of Chile Sebastián Piñera and spans more than 1180 km^2 on the southwestern side of the island.

Ancud and Castro are the largest cities on the archipelago, each with a population of about 40,000. Along with Quellón to the south, they have grown considerably since the 1990s as aquaculture and tourism have come to form the largest economic sectors in Chiloé. During this period, the archipelago has undergone a historic demographic shift, from a majority rural population to one that is more than 60% urban today. This shift has had significant consequences on the livelihoods and basic economic arrangements of islanders. Artisanal fishing and subsistence farming, though still practiced, are less common, as are the communal labor traditions that accompanied these livelihoods. Most Chilotes today depend on wage labor in both the private and public sectors, often working long hours for relatively low pay. Many of these jobs are unreliable from season to season. The fishing and aquaculture industries are particularly vulnerable to fluctuations in climate and ecosystem, as was evidenced by the devastating red tide (*marea roja*) outbreak of 2016, when Chile's national government placed a ban on sales of fish and shellfish that had been contaminated by toxic algae blooms. Economic activity on the archipelago came to a near standstill, and thousands of Chilotes lost their jobs. Indeed, the phenomenon of toxic algae blooms has become a recurring nightmare for islanders in recent years as warming ocean temperatures (a consequence of climate change and El Niño

---

[1] *Greigia sphacelata*

[2] *Ulex europaeus*

[3] The most common species of trees in Chiloé's old growth forests are *Metrosideros stipularis* (a type of myrtle tree known locally as *tepú*), *Nothofagus dombeyi* (called *coigüe*), *Aextoxicon punctatum* (*olvillo* or *tique*), and *Luma apiculata* (*arrayán*).

weather patterns) and contamination from agricultural runoff and salmon farms exacerbate the problem.

With a population of more than 168,000,[4] Chiloé is classified as a province. Until the mid-twentieth century, the administrative boundaries of the province included various sectors of the mainland that maintained strong commercial ties with towns on the archipelago. Those mainland sectors, located across the Sea of Chiloé along the northern boundary of Patagonia, are often referred to informally today as Continental Chiloé and they remain culturally similar to the archipelago. Our ethnobiological overview of Chiloé extends to this part of Chile, and even further south, to include the port city of Aysén and its surroundings. Although part of the mainland, these settlements developed in relative isolation, much like settlements on the islands of Chiloé, because of their remoteness from Chile's larger cities to the north. Indeed, until construction of what is generously termed a "highway"—the Carretera Austral—in the 1980s, Aysén was accessible only by sea, or by roads that originated in Argentina.

The History

Social scientists have long recognized that no place develops in total isolation no matter how geographically remote it may be. The Archipelago of Chiloé is no exception. Over the centuries, people have migrated to and from the islands, sometimes in large numbers, sometimes in a trickle. They have introduced new ideas, species, and technologies that have altered to varying degrees the relationship between islanders and their environment. As with any place, that history is reflected in today's social and ecological landscape. As noted by Paul E. Minnis, "The present has a past, which lives on in us, ecologically, culturally, and politically" (2016:xi).

The most accelerated changes in Chiloé occurred during two periods: early colonization in the late 1500s, and industrialization during the latter half of the twentieth century. Both of these periods were characterized by new arrivals and the development of new life ways. Nevertheless, certain practices and traditions endured, remaining remarkably constant over the centuries. I will present here a general overview of Chiloé's last 500 years of history, calling attention to some of the most significant elements of change and continuity.

[4] Based on Chile's 2017 census.

Indigenous Origins

Prior to contact with European explorers, two indigenous groups occupied the islands of Chiloé. The Chonos were a nomadic, seafaring Patagonian group that migrated north from the Archipelago of the Chonos and the mainland more than 2000 years ago. They traveled in portable canoes called *dalcas* and relied on a diet of fish, shellfish, potatoes, and, on occasion, larger fauna like sea lions. They often gathered shellfish in small bands, cultivated seaweed, and built mud-and-stick barriers in shallow estuaries to trap fish at low tide, a technique (*corrales de pesca*) commonly used in Chiloé until the late twentieth century. Their dalcas, ideally built for navigation across estuaries and narrow ocean channels and easily carried across land, remained the principal form of transportation for islanders until the late nineteenth century and were still in use as late as the 1930s.

Europeans first came into fleeting contact with these islanders in the mid-sixteenth century. Spanish explorer Juan Fernández Ladrillero traded with Chono bands in 1557. One of his crewmembers, Miguel de Goicueta, wrote that they were "of average height," covered in brown clay and sea lion skins, and carried wooden spears and "daggers of whale bone." He observed them diving out of their dalcas into the frigid ocean waters in search of shellfish (Goicueta 1880:410). John Byron, a junior officer in the British navy who survived a 1740 shipwreck with the help of Chonos, described them as "amphibious in nature," noting "the sea is the only source from whence almost all their subsistence is derived" (Byron 1983:214,220). He observed them snaring birds, gathering shellfish, and fishing with techniques that were still in practice 200 years later. Although the Chonos ceased to exist as a distinct ethnic group sometime in the nineteenth century, aspects of their culture have persisted to the present (Cárdenas et al. 1991, Urbina Burgos 2004, Álvarez Abel 2012).

The second indigenous occupants of Chiloé—more numerous than the Chonos—were the Huilliche, a sedentary, agricultural group that migrated to the archipelago from south-central Chile sometime before the fifteenth century. An offshoot of Chile's largest indigenous group—the Mapuche—the Huilliche spoke the same language (Mapudungun) as the Mapuche, lived in scattered villages governed by a headman, or *lonko*, and practiced subsistence horticulture, hunting, fishing, and some gathering. Their most important crop was the potato, a food native to Chiloé, of which they cultivated hundreds of varieties.

The Huilliche engaged in a form of labor sharing called the *minga*, a practice that was common among sedentary indigenous groups throughout the Americas prior to contact. Mingas took one of two forms: large communal work parties for particular projects, like the building of a home, or reciprocal labor arrangements between neighbors. Because of the geographic isolation of many island communities in Chiloé and the absence of a cash economy, mingas remained widespread well into the twentieth century, becoming a source of cultural pride and a tradition emblematic of islander identity. They are still practiced in some of the more rural and remote island communities in Chiloé where the cash economy remains weak.

Colonial Years

The arrival of Spanish settlers to Chiloé in 1567 marked a significant turning point in the relationship islanders had with their natural environment. These first colonists, numbering less than 200, introduced new crops, livestock, and technologies that merged with existing indigenous practices to produce a uniquely Chilote way of life. Led by Martín Ruiz de Gamboa, they established outposts at Chacao and Castro, hoping to exploit the Huilliche population for labor and to find gold. Instead, during the first years of colonization, their ranks thinned and their supply chains dwindled, and they found themselves largely dependent on the local indigenous groups for survival.

That dependence deepened in 1598 when Mapuche groups on the mainland north of the archipelago rebelled against Spanish colonists and reclaimed much of southern Chile. Although Huilliche groups in Chiloé did not take part in the rebellion, the effect was to further isolate settlers in Chiloé from Spanish-controlled territory to the north. For the next two centuries, Chiloé remained a poor, politically marginalized province. From the perspective of Spanish administrators in Santiago and Valparaíso, it offered no significant resources and served primarily as a stopping point for ships heading to and from the Straits of Magellan. Castro, the principal settlement in Chiloé, was described by one official as "a recently inhabited city of little value and with no gold" (Cisneros 1956: 380). A supply ship was sent to Chiloé from the seat of regional colonial power—the Viceroyalty of Peru—every 1 or 2 years. Beyond that, islanders were offered little in the way of assistance from the larger settlements in Chile's central valley.

Throughout the colonial period, Chilotes continued to rely on small-scale agriculture and maritime activities like fishing and shellfish gathering for subsistence. However, they also incorporated the new plant and animal species that had been introduced by Spanish settlers.

The earliest evidence of Old World livestock in Chiloé comes from contemporaneous accounts of Martín Ruiz de Gamboa's 1567 expedition to the Isla Grande. With the help of Huilliche bands, the colonists crossed the two-mile channel of Chacao on dalcas with several hundred horses swimming alongside.[5] They brought fowl with them aboard the dalcas, and within a few years, had pigs, cattle, and sheep. These animals became a mainstay of subsistence, assisting islanders with their agriculture and transportation, and providing a limited supply of meat and wool.

[5] Two letters from 1567 describe this dramatic crossing to Chiloé's Isla Grande: a May 12 letter from the cabildo of Osorno and a November 20 letter written by officials in Concepción (see *Colección de documentos inéditos, segunda serie, tomo I, 1558–1572* [Santiago: Fondo Histórico y Bibliográfico J.T. Medina] 1956, p.92 and p.104). Most historians who have written on the subject accept the figure of 300 as the number of horses that swam across the channel; however, Diego Barros Arana (*Historia general, tomo II* [Santiago: Rafael Jover] 1884 p.369, f15) claims that only 15 horses were brought to Chiloé on the expedition. Alonso de Góngora Marmolejo, writing in the 1570s, also describes the event (in *Historia de Chile desde su descubrimiento hasta el año 1575* [Santiago: Editorial Universitaria, S.A.] 1969 [c1575], p.98).

Potatoes continued to be the principal crop farmed by colonists. But wheat, rye, and oats were introduced as supplementary crops in the sixteenth century. Other Old World species that responded favorably to the climate of Chiloé and were widely incorporated into local diet were apples, onions, garlic, mint, clover, and carrots. Apples came to replace local fruits like strawberries, *molle*, and Calafate berries as the most common source of *chicha* (fermented cider) (Daughters 2014; see also Chap. 4).

The merging of Spanish and indigenous practices extended to nearly all aspects of cultural life on the archipelago. In the immediate aftermath of the founding of Castro in 1567, Huilliche groups living alongside the new settlement assisted the Spaniards in the building of homes, a modest plaza, and a small church. While Spaniards imposed the centralized city layout that characterized most of their New World settlements, construction techniques were in many cases borrowed from the Huilliche, with Spanish dwellings coming to resemble some of the larger Huilliche structures already in place (Urbina Burgos 2004). Like those of the Indians, settlers' houses were built with logs, crudely shaped planks, or branches, with gaps in the walls covered by "pieces of animal hides and old rags" (González de Agueros 1791: 111).[6] The roofs were made of straw, and the floors were of bare earth. A *fogón*, or fire pit, occupied the center of the house, continually filling the single-room dwellings with smoke and causing "infinite discomfort" to the residents while preserving the straw roofs from decay (Beranger 1893: 20).[7] Despite the haze, wrote one observer, "everything is within sight upon entering, [including] chickens and other domesticated animals" (González de Agueros 1791: 111).[8]

The borrowing of indigenous materials and techniques held true as well for the construction of sea vessels. Spanish settlers quickly abandoned their cumbersome rowboats and dinghies in favor of the more maneuverable and portable dalcas. Dalcas could be built entirely out of local materials, meaning that iron nails—a commodity that was extremely hard to come by throughout much of the archipelago's colonial history—were unnecessary. The local vessel, furthermore, was better suited to navigate quickly and efficiently the choppy waters that separated the neighboring islands, having the added advantage of being collapsible for brief sojourns over land. In a region devoid of roads, dalcas, in short time, became the predominant vehicle of Chiloé.

The Spaniards also adopted the most common and efficient subsistence strategies employed by the Huilliches. Fish corrals became a regular source of food, as did the archipelago's extensive shorelines from which shellfish, algae, and other resources were easily collected. Agriculture was widely practiced, but under the same circumstances—and using many of the same techniques—of the Indians. The absence of large tracts of cleared land on the islands precluded the establishment of extensive haciendas, such as those found in central Chile. It also curtailed the

[6] "*...pellejos de carnero y trapos viejos...*"

[7] "*...cuyo humo que deben sufrir a costas de infinitas incomodidades...*"

[8] "*Luego que se entra del umbral de la puerta para adentro, está a la vista toda la casa con cuanto en ella tienen, y allí se hallan también las gallinas y otros animales domésticos.*"

practice of large-scale agriculture, even among those soldiers and settlers who were apportioned grants of Indian labor (*encomiendas*) by Ruiz de Gamboa. Instead, the majority of Spanish colonists, much like the indigenous inhabitants, worked smaller, individually owned plots of land—an arrangement of land-use and land-distribution that would eventually morph into the ubiquitous *minifundio* ("small farm") of nineteenth- and twentieth-century Chiloé.

The most common agricultural crop farmed by the first colonists was a root native to the archipelago: the potato. More than 200 subspecies of the tuber, cultivated by local bands for several millennia, could be found throughout the islands. Potatoes were quickly incorporated into the colonists' diet and shipped to settlements on the mainland. They were, moreover, Chiloé's first *global* export. Picked up by Sir Francis Drake during an encounter with Huilliches in southern Chile in 1578, a half a dozen varieties of the crop were taken to Ireland and spread across Europe.[9]

Given that most plots of land were painstakingly tilled by hand by women and men, farming techniques were borrowed from the Huilliche as well. The usual method was simple but labor intensive. Mounds of earth were overturned by shoving two heavy wooden poles into the ground "with a thrust from the hip," as described by one missionary from the period. "In this manner they cover the entire field, suffering all the while from a savage fatigue" (Hanisch 1982: 245).[10] Those Spaniards awarded encomiendas were, in the early years of colonization, able to rely on Indian labor for these tasks; however, the absence of a cash economy on the archipelago, along with the inexorable decline of the institution of the encomienda, paved the way for the widespread adoption of indigenous reciprocal-labor practices among the Spanish colonists. As late as the mid-twentieth century, the *minga* was still the predominant means by which agricultural work was accomplished.

Chiloé under Chile

These practices lasted centuries in Chiloé. Because of the relative isolation of the region, change was slow. When Chiloé was forcibly integrated into the new nation of Chile in 1826 following fierce resistance by islanders and a series of battles on the archipelago, Chilotes found themselves subsumed into an entirely different political structure, beholden to the governing body of a nation-state headquartered in Santiago rather than colonial administrators in Lima. But in their day-to-day subsistence—their relationship to the broader natural world—life changed very little. They fished, they farmed; they tended their livestock; they gathered shellfish and seaweed. They engaged in barter, relied on their neighbors for labor, drank chicha, played music, and told stories at night in the fogón.

---

[9]All varieties of potatoes found throughout the world today originated from two places: Andean South America (Peru and Bolivia) and Chiloé. See Chap. 4 of this volume.

[10] *"Y así prosiguen todo el terreno con una bárbara fatiga."*

When Charles Darwin visited Chiloé in 1834, he was struck by the remoteness and poverty of the archipelago's settlements. He noted that in Castro he was "unable anywhere to purchase either a pound of sugar or an ordinary knife. No individual possessed either a watch or clock; and an old man, who was supposed to have a good idea of time, was employed to strike the church bell by guess" (Darwin 1998: 63). He described the forests as impenetrable, such that "the land is nowhere cultivated except near the coast and on the adjoining islets. Even where paths exist, they are scarcely passable" (Darwin 1998: 57).

Regional economic changes in the late nineteenth and early twentieth centuries eased some of these hardships and began to lift the isolation experienced by islanders. A boom in forestry exports from southern Chile in the 1890s drew some Chilotes to wage labor jobs on the mainland and increased the circulation of cash in Chiloé through remittances sent to families on the archipelago (Klubock 2014). The increased availability of machine-cut wood also changed the shape and style of Chilote houses to the wood-tile look characteristic of the twentieth century. The construction of a narrow-gage rail line in 1911 dramatically reduced travel time between the two largest towns on the Isla Grande—Castro and Ancud. Although destroyed by the earthquake of 1960, the rail line provided one of the most reliable and secure—albeit slow—modes of land transportation through the mid-1900s.

Nevertheless, transportation by boat remained more widespread throughout the nineteenth and most of the twentieth centuries. By the 1920s, locally built sailboats had replaced dalcas as the most commonly used vessel. The first boat engines were imported in the 1950s, expanding the range of fishermen, reducing travel time between island communities, and creating a limited dependence on gasoline. And Japanese-made nylon fishing nets, introduced the following decade, ushered in what was referred to locally as *el milagro de las redes* (the miracle of the nets) because of their better efficiency and durability.

Overall, the flow of goods to and from Chiloé increased, easing some of the economic isolation of Chilotes and diversifying their livelihood options. Yet most of Chilote society remained rural. The archipelago's largest towns—Castro and Ancud—each had populations of about 4000 in 1940 (Instituto Nacional de Estadísticas n.d.). Cash circulated throughout the islands, but a majority of islanders still practiced subsistence farming, fishing, and livestock-raising, drawing from both indigenous and Spanish traditions, as they had for centuries.

Early Industrialization: Fishing and Aquaculture

The most accelerated changes to islander life took place during the latter half of the twentieth century. In 1956, Chile's national government designated Castro as a duty free zone, diversifying commerce throughout the city (Jiménez 2009). With transportation to the mainland increasingly affordable to islanders, many men and women opted to travel to Patagonia and southern Argentina for jobs as shepherds, urban laborers, domestic servants, and shopkeepers, sending money back to families on the archipelago. By the 1960s, the economies of Chiloé's largest towns were based more on market exchange than reciprocity. Commercialism was beginning to take hold.

The shift in Chiloé from a rural agrarian economy to an industrialized one gained momentum following the coup d'etat that toppled Salvador Allende in 1973. Under a right-wing military junta headed by General Augusto Pinochet, Chile adopted a series of neoliberal market reforms that opened up the country's natural resources to large extractive industries. Private business was promoted as the motor for economic growth: trade barriers were struck down, regulations rolled back, and organized labor systematically weakened. Nationally, exports of fruits and vegetables, lumber, and seafood all increased during this period.

Chiloé underwent a boom in shellfish and seaweed exports in the 1970s and 1980s, drawing workers from the mainland to the archipelago and accelerating the economic shift to wage labor. Extraction of Chilean abalone—known as *loco*—reached a zenith in the early 1980s in response to a surge in demand from Japan (Schurman 1996). By the late 1980s, wild stocks of the shellfish had been largely depleted, and in 1989, the Pinochet government was forced to abandon its anti-regulation posture and impose a total ban on loco harvesting, a prohibition that was recently renewed and extended through the year 2022 by Chile's National Fisheries and Aquaculture Service (Sernapesca).

The seaweed *pelillo* (*Gracilaria chilensis*)—a grassy substance that grows wild along the shallow coastal waters of the archipelago—underwent a similar pattern of extraction and depletion. Discovered to contain a key ingredient for common food preservatives, pelillo became a significant regional export in the 1960s. Islanders would gather wild pelillo along the beaches and even harvest their own just offshore of their properties, selling bundles of the seaweed to distributors. Exports peaked in the 1980s but dropped off precipitously by 1990 as the resource dwindled. Since then it has rebounded as an export product and today is a mainstay of small-scale aquaculture in Chiloé.

The fishing sector also expanded significantly during this period. Small-scale artisanal fishing—*la pesca artesanal*—became a relatively lucrative venture for some islanders. Large-scale Chilean and foreign fishing operations also ventured into the waters surrounding the archipelago. Extraction efforts targeted wild stocks of southern hake, sea bass, and conger eel for export to Spain, the United States, Japan, and mainland Chile. Fish exports soared in the early 1980s, but by the end of the decade, all three species were showing signs of decline. Southern hake (*merluza del sur*) was hit especially hard, with stocks in southern Chile falling 66% between 1981 and 1991 (CORFO/IFOP 1991, Schurman 1996).

The Rise of Salmon Farms

As stocks of wild fish were dropping and shellfish exports were leveling off in the late 1980s, a new industry was launched that by the end of the century would boast record-setting growth: farm-raised salmon.

Southern Chile's ocean temperatures and coastal geography provided an ideal environment in which to raise salmon. These favorable conditions, coupled with the availability of cheap labor and lax government oversight, attracted national and foreign aquaculture companies to the archipelago (Barton and Fløysand 2010). Several modest public and private aquaculture ventures in the 1970s—including the introduction of Chinook salmon to Curaco de Vélez, Chiloé—showed that salmon farming in southern Chile was viable and potentially lucrative (Basulto 2003; United Nations 2006). However, the industry really took off in the 1990s when hundreds of farms sprung up along Chiloé's inland lakes—where the fish were grown for the first months of their life—and eastern fiords—where they were fattened and harvested for export. The three most commonly farmed species were Atlantic salmon, Coho (Pacific salmon), and Rainbow trout. Between 1987 and 1996, production levels grew by more than 1800%, and by 2001, the farmed species accounted for 5% of Chile's export earnings, making Chile the second-leading exporter of salmon in the world (Pitchon 2015; Schurman 2003).

Production reached a zenith in 2006. That year, 47 companies operated more than 500 farms in southern Chile (Revista Salmonicultura 2006:19). The industry employed more than 45,000 people directly and indirectly in the Province of Chiloé and the larger Lakes Region (Región de Los Lagos) (United Nations 2006: 26).

The dizzying rise of aquaculture came at an ecological cost. Salmon farms disgorged intense concentrations of feed, antibiotics, and salmon feces, polluting lakes and coastlines. With hundreds of thousands of salmon often packed into farms no larger than football fields, the surrounding waters sometimes turned into oxygen-depleted "dead zones" within 2 or 3 years. Storms often tore nets at salmon farms, causing hundreds or thousands of fish to escape at a time.[11] Most of these escapees were Atlantic salmon—a nonnative species. The effect of these foreign carnivores on the local ecosystem is still poorly understood (Image 1.1).

Urbanization

In the 1990s, Chiloé's social landscape began to undergo a fundamental shift as its population became more urban and proletariat. With agricultural activity becoming increasingly untenable for small-scale producers, many rural Chilotes took up jobs at processing plants—the large, refrigerated warehouses where fish and shellfish were prepared for shipment to Chile and beyond. Others were employed full-time at the hundreds of salmon farms that dotted the archipelago's lakes, inlets, and fiords. Meanwhile, jobless Chileans from the mainland continued migrating to the larger towns of Quellón, Ancud, Castro, and Dalcahue in search of employment in processing plants, canneries, and salmon farms. This latter development was a new phenomenon for Chiloé, which until the 1970s had experienced a net *outward* migration of Chilotes to the mainland.

[11] Interview with Juan Carlos Cárdenas, Santiago, March 2004.

Image 1.1 Artisanal fisherman near Dalcahue protest encroachment of salmon farms on their fishing territory in 2006. (Photo credit: Anton Daughters)

Altogether, these population shifts significantly altered the archipelago's human landscape. The total population of the archipelago shot up, increasing by nearly 50% between 1970 and 2002 (from 105,000 to 154,000) (INE n.d.). Nearly all of this growth took place in Chiloé's largest towns. Quellón, Ancud, and Castro doubled in size within two decades.

Of particular significance during this time was the fact that the archipelago's *rural* population decreased considerably. While nearly three-quarters of all islanders lived on small farms (minifundios) in 1970, that figure had dropped to 52% by 1992. By the mid-1990s, Chiloé had crossed a milestone: for the first time in the archipelago's history, rural Chilotes made up less than half of the population.

Today the number of rural islanders stands at 39%. This overall decline, moreover, is reflected not just in the percentage of the population but also in real numbers—from 72,798 in 1970 to an estimated 66, 694 in 2012 (INE n.d.).[12] In other words, despite a significant increase in the overall population of the archipelago in recent decades, there are fewer rural islanders today than in 1970.

[12] This is based on the population numbers of Ancud, Castro, and Quellón in the 2012 census. It is worth noting that the methodology of the census was challenged and the results declared ineligible. However, they are the most recent counts of the population of those cities (the 2017 census gave total population at the national, regional, and communal levels only).

Chiloé Today

Chilotes are not alone in navigating big economic changes. Nearly all communities around the world have been affected in one form or another by the forces of globalization and industrialization. What makes Chiloé somewhat unique is the speed with which these changes have occurred. Within a period of about 40 years—from the 1950s to the 1990s—its economy shifted from one based primarily on subsistence fishing and farming to one of wage labor and global exports.

The memory of that rural past is strong, and islanders speak of it with reverence. When describing their broader cultural identity, they often discuss *mingas* and minifundios, fishing and farming, shellfish-gathering with their *tía* or *tío*, collecting seaweed by the shore, or participating in *curantos*.[13] They invoke these rural activities even when they have grown up in an urban setting with internet, iPhones, and satellite TV (Daughters 2016b). They speak with pride of their reliance on the ocean and the earth, and of their traditions of reciprocity and barter. And they often reserve their harshest criticism for what they see as the biggest symbols of so-called progress and modernity: the salmon-farm industry, the proposed bridge to the Isla Grande, and the recently built mall of Castro (Miller 2016).

The biggest source of employment in Chiloé today is the aquaculture industry. But that sector has faced serious challenges in recent years. The industry suffered a near collapse with the outbreak of the infectious salmon anemia virus (ISA) in 2007. Over a 3 year period, the disease wiped out most stocks of Atlantic salmon, with production levels dropping from 400,000 tons in 2008 to 100,000 tons in 2010 (Alvial et al. 2012). Biologists had long warned that poor sanitation control and the overcrowding of salmon in farms—a consequence of lax regulations—could lead to significant losses in the event of a viral outbreak. Those premonitions proved true, and the losses of 2008–2010 are the largest the industry has suffered since its inception in southern Chile in the mid-twentieth century.

In addition to production losses, more than 25,000 jobs (about half of all direct and indirect industry jobs) were lost in the Lakes Region from 2008 to 2010, more than 10,000 of those in Chiloé alone (Alvial et al. 2012: 14). Some unemployed islanders returned to their rural properties during the crisis, and there was a brief uptick in subsistence fishing and agricultural production in 2011. Meanwhile, the aquaculture industry shifted its production south, opening new farms in Aysén. Production levels of farmed salmon in the region have slowly climbed again and by 2016 were back to pre-crisis levels.

[13] A *curanto* is a traditional meal of fish, shellfish, potato patties, and chicken, layered by *nalca* leaves and cooked—in a pot or in a pit in the earth—for several hours. Curanto meals are often big social occasions in which neighbors and extended family gather to help prepare the food (see Chap. 4 of this volume).

The Crisis of 2016

Whereas the setbacks to the salmon industry in 2008–2010 had lasting social consequences in Chiloé, the crisis of 2016 mobilized islanders on a scale that had never been seen before. Chilotes found their livelihoods threatened on several fronts: massive toxic algal blooms in the ocean, contamination by the salmon industry, and a government ban on the production and consumption of fish and shellfish that left their economy paralyzed.

Signs of trouble in the waters surrounding the archipelago began in late 2015 when hundreds of dead whales and seagulls mysteriously washed up on the shores of Patagonia and Chiloé. In the months that followed, the pace of marine die-offs quickened, with tens of thousands of dead sardines and millions of contaminated mollusks clogging river deltas and beaches from Valdivia to Chiloé. The salmon industry was hit in March of 2016 when more than 23 million farmed salmon died from a natural toxin in the waters north of Chiloé's big island (Esposito 2016, Pfeiffer 2016). By April, the region was in a full-blown environmental crisis. Islanders spoke of being "in anguish" over the state of the ocean, of feeling like they were "navigating over a toxic pond."[14]

There is still disagreement over the full range of causes of the die-offs. Most biologists point to the most immediate cause—the historically unprecedented high levels of toxic algal blooms that year. Known as *marea roja* (red tide), these algal blooms can poison and choke off marine life in large swaths of the ocean. Unusually high water temperatures in 2016—the result of the El Niño weather phenomenon and human-caused climate change—led to the most severe red tide outbreak on record for that sector of the Pacific. A contributing factor may also have been contamination from the more than 400 salmon farms in the region. The intense concentration of fishmeal, feces, and antibiotics produced at salmon farms are known to spur the growth of micro algae, accelerating the eruption of toxic red tides under certain conditions (Buschmann et al. 2006, Sellner et al. 2003) (Image 1.2).

Many Chilotes, however, suspect that the salmon industry played a larger role in the die-offs than government and industry officials have claimed (Franklin 2016, Pfeiffer 2016; also, see Chap. 3 of this volume). In March of 2016, salmon farm companies dumped thousands of tons of rotting salmon into the waters northwest of Chiloé's big island. The environmental NGO Greenpeace argued that this waste, which would have produced a massive dead zone in the ocean, may have been pushed landward by currents, choking off fish in its wake (Soberanes and Pérez 2016). The claim has plausibility given that many of the fish-kills occurred *after* the salmon dump and north of the archipelago.

When dozens of people fell sick after eating shellfish contaminated by the algal toxins, the government of Chile placed a ban on the production and consumption of fish and shellfish. Those directly affected by the ban were promised government monthly payments of US$150 (100,000 pesos) for 3 months.

[14] Phone interview with Daniela Leviñanco, April 15, 2016.

Image 1.2 Graffiti in Ancud in May 2016 reflecting anger toward salmon farms in Chiloé. (Photo credit: Anton Daughters)

Islanders erupted in fury. They were in an impossible bind, they argued. Their livelihoods were cut off, and the offers of compensation from the government were entirely inadequate. Moreover, the government's Nation Fishing Service—Sernapesca—was shielding the salmon-farm industry by not discussing the full range of possible causes behind the crisis. In early May of 2016, protesters blockaded the principal access ramps to Chiloé's big island, effectively sealing off the archipelago from the outside world. The burning barricades and motley crowds of protesters—made up of fishermen, truck drivers, scuba divers, shopkeepers, service workers, and high school students—caught the attention of the international media. Reports of the crisis made headlines in major newspapers around the world.

Under the rallying cry of "*Chiloé está privao!*" ("Chiloé has been deprived"), islanders organized massive peaceful marches in Castro, Quellón, Ancud, and Chacao. Chilotes from all walks of life spoke at the rallies, describing how the crisis impacted them personally, sometimes commenting on the broader struggles their families had faced over the years as a consequence of rapid economic changes.[15] Among the demands of the protesters: tougher environmental regulation of the salmon industry, a reshaping of national fishing laws to facilitate subsistence and small-scale commercial fishing, and greater monetary recompense for affected families (Image 1.3).

---

[15] Richard Vercoe, living on Chiloé's big island during the crisis, described these rallies in a phone interview on May 11, 2016.

Image 1.3 A flyer in Dalcahue promotes a rally in May 2016. (Photo credit: Anton Daughters)

Leaders of local artisanal fishermen syndicates and scuba diver unions negotiated with government officials eager to bring an end to the standoff. President Michelle Bachelet appointed Chile's economic minister Luis Felipe Cespedes to oversee the negotiations. In late May, the government agreed to triple the amount of compensation offered to islanders and pledged to work toward stiffening oversight of salmon farms. Some islanders considered the government concessions inadequate—empty promises absent of real policy change. But with supplies of basic goods dwindling throughout the archipelago and the threat of severe privation becoming a reality, most union leaders agreed to end the protests. Access ports were reopened, and islanders began the long process of reconstituting their lives and livelihoods.

The onset of winter in June of 2016 brought lower ocean temperatures, dissipating the red tide phenomenon. But islanders remained wary, anxious that the problem would return, perhaps with greater severity, within a few years.

Challenges Ahead

The crisis of 2016 highlights the biggest challenges that will be faced by islanders in the years ahead. The first and most immediate of these is how to contend with the growing phenomenon of red tide that threatens to close off the ocean as a source of sustenance. With average ocean temperatures projected to rise over the next decade (Laffoley and Baxter 2016, Wallace 2016), islanders will almost surely have to grapple with more—and more massive—outbreaks of harmful algal blooms (HAB). This is a grim prospect for a people who have built a culture on ocean-based livelihoods.

Three factors contribute to the growing red tide phenomenon: (1) El Niño, (2) climate change, and (3) contamination from fish farms and agricultural runoff. Little can be done to address the first of these. The weather event known as El Niño occurs at irregular intervals of between 2 and 7 years and can last up to 2 years, warming the waters off the coast of South America and thus creating opportune conditions for harmful algal blooms (NOAA). However, the problems of climate change and contamination from aquaculture and agriculture *can* be addressed, perhaps limiting the severity of future HAB outbreaks. With regard to climate change, Chile has taken steps in recent years to reduce its carbon emissions, signing the Paris Accord and pledging in early 2018 to shift to a majority of renewable electricity-generating sources (wind, solar, geothermal, biomass, and hydro) by 2050. Islanders can continue to pressure policy makers in Valparaíso and Santiago, holding them to task on the pledges they have made, and pushing for more aggressive action against climate change.

The task of strengthening government oversight of the salmon industry has proved to be, perhaps not surprisingly, much tougher. For decades, union leaders, local activists, scientists, NGOs, and academics have been pointing to the damage fish farms cause the surrounding ocean habitat. That damage can be minimal if salmon aquaculture is carried out within certain constraints. For example, limiting the number of fish packed into farms, the amount of antibiotics used, and the number of farms per sector can lead to relatively low-impact aquaculture operations. Most environmentalists—and islanders—agree that the regulations in place today are insufficient. But regulations with real teeth are difficult to pass because of the close relationship between the salmon-farm industry and the regulatory arms of the Chilean government—Sernapesca and Subpesca (agencies that are under the Ministry of Economy).

The revolving door between the private and public sectors in Chile is real and likely the biggest obstacle to regulatory change in the salmon industry. For example, following his term as the head of Subpesca, where he oversaw policy of Chile's fishing sector, Felipe Sandoval joined the board of AquaChile, one of the largest salmon growers in Chile, in 2010, and was appointed in 2013 to head SalmonChile—the largest consortium of salmon-farm companies in Chile. In 2016, conservative parliamentarian Marta Isasi was placed under house arrest for allegedly accepting payments of 25 million pesos (more than US$52,000) by Corpesca, one of Chile's biggest producers of fishmeal, in anticipation of a vote on the 2012 *Ley de Pesca* (the current national fishing law). These and other examples help explain why the government has not taken more aggressive action in the wake of the 2016 crisis.

Nevertheless, the 2016 crisis made one thing clear: the degree to which islanders of all walks of life are dependent, either directly or indirectly, on the ocean. And the health of that ocean very much depends on the degree to which the Chilean government exercises oversight of its marine resources. In this regard, politics, ecology, economics, and culture are deeply intertwined in Chiloé.

Adriana Yañez, born and raised in Chiloé to parents who were fishermen, gave voice to these concerns during a rally in Ancud on May 10, 2016. Her words reflect the complexity of the challenges islanders face today:

> The problem is more than just red tide outbreaks. It has to do with a system that favors large export industries at the expense of our livelihoods, our heritage, and our sense of who we are.... We have gone from maritime hunters and gatherers to people who can no longer depend on the sea for sustenance. How is it that we have allowed a privileged few who have no idea what life in the ocean is like to destroy our sea? ...But enough with so much restlessness. We all carry some responsibility. Let's stop waiting for politicians to solve the problems in our lives, let's solve our own problems by letting them know once and for all that they work for us.[16]

Organization of the Book

This volume is by no means an exhaustive account of the ethnobiology of Chiloé. As with any anthology, in choosing to focus on certain topics we are, by necessity, omitting others. We hope that the strength of the research presented in the following chapters makes up for any of those omissions.

The chapters are grouped thematically, with Chap. 2 (by Ana Pitchon) and Chap. 3 (by Sarah A. Ebel) examining ocean-based livelihoods (aquaculture and fishing), and Chap. 5 (by Richard Vercoe) and 6 (by Eric H. Thomas) focusing on land-based resources (potatoes and wood) as well as trade (the *trueque* Chilote). Chap. 4 (by Anton Daughters) offers a bridge between marine and land subsistence traditions by describing the history and cultural significance of potatoes, *curanto*, and chicha. And Chap. 7 (by Giovanna Bacchiddu) offers a detailed ethnographic case study of the formation of a rural household on the island of Apiao. All of the research presented here is based on participant observation fieldwork in Chiloé and Patagonia carried out, in most cases, over several years. The chapters, moreover, incorporate both qualitative and quantitative methods. Little has been published in English about Chiloé and its environs, even less about the ethnobiology of the archipelago. Altogether, this volume offers the most comprehensive ensemble of original research examining human-nature relationships in this region of Chile.

References

Álvarez Abel, Ricardo. 2012. Prácticas Rituales Asociadas a Tierra y Mar: Quepucas y Treputo. In *Chiloé: Historia del Contacto*. Gráfica LOM: Santiago de Chile.

Alvial, Adolfo, Frederick Kibenge, John Forster, José M. Burgos, Rolando Ibarra, and Sophie St-Hilaire. 2012. *The Recovery of the Chilean Salmon Industry: The ISA crisis, its consequences and lessons*. Puerto Montt: Global Aquaculture Alliance Feb. 23. https://www.aquaculturealliance.org/wp-content/uploads/2015/04/goal11-adolfoalvial.pdf.

Barton, Jonathan, and Arnt Fløysand. 2010. The political ecology of Chilean Salmon aquaculture, 1982-2010: A trajectory from economic development to global sustainability. *Global Environmental Change* 20 (4): 739–752.

[16] Quoted in Daughters 2016a.

Basulto, del Campo Sergio. 2003. El largo viaje de los salmones: una crónica olvidada. In *Propagación y cultivo de especies acuáticas en Chile*. Editorial Maval: Santiago de Chile.

Buschmann, Alejandro, Verónica A. Riquelme, María C. Hernández-González, Daniel Varela, Jaime E. Jiménez, Luis A. Henríquez, Pedro A. Vergara, Ricardo Guíñez, and Luis Filún. 2006. A review of the impacts of salmonid farming on marine coastal ecosystems in the Southeast Pacific. *ICES Journal of Marine Science* 63 (7): 1338–1345 https://doi.org/10.1016/j.icesjms.2006.04.021.

Byron, John. 1983. In *The Narrative of the Honorable John Byron (1768)*, ed. Christopher Hibbert. London: The Folio Society.

Carlos de Beranger. 1893 [1773]. Relación geográfica de la provincia de Chiloé. Santiago de Chile: Imprenta Cervantes.

Cárdenas, Álvarez Renato, Montiel Vera Dante, and Grace Hall Catherine. 1991. *Los Chonos y los Veliche de Chiloé*. Santiago de Chile: Ediciones Olimpho.

CORFO/IFOP. 1991. *Diagnóstico de las principales pesquerías nacionales: estado de situación y perspectivas del recurso pesquerías demersales 'peces' zona sur austral, 1990*. Santiago de Chile: CORFO/IFOP.

de Cisneros, Agustín. 1956. Declaración del licenciado Agustín de Cisneros sobre el monto de los diezmos del obispado de la Imperial, September 1, 1571. In *Colección de documentos inéditos, segunda serie, tomo I*, 1558–1572. Santiago: Fondo Histórico y Bibliográfico J.T. Medina.

de Goicueta, Miguel. 1880. Viaje del Capitán Juan Ladrillero (1557-1558) al Descubrimiento del Estrecho de Magallanes. *Anuario Hidrográfico de la Marina de Chile* 6: 482–520.

Darwin, Charles. 1998. *Chiloé*. Santiago de Chile: Editorial Universitaria.

Daughters, Anton. 2016a. Fish kills and protests on the islands of Chiloé. *Anthropology News* 57: 61–66.

———. 2016b. Southern Chile's Archipelago of Chiloé: Shifting identities in a new economy. *Journal of Latin American and Caribbean Anthropology* 21 (2): 317–335.

———. 2014. Of Chicha, Majas, and Mingas: Hard apple cider and local solidarity in twenty-first century Rural Southern Chile. In *Alcohol in Latin America: A Social and Cultural History*. Tucson: University of Arizona Press.

Esposito, Anthony. 2016. Chile's Salmon Farms Losing up to $800 Million from Algal Bloom. *Reuters*. March 9.

Franklin, Jonathan. 2016. Toxic red tide in Chile prompts investigation of salmon farming. *The Guardian*. May 17. https://www.theguardian.com/world/2016/may/17/chile-red-tide-salmon-farming-neurotoxin

González de Agueros, Pedro. 1791. *Descripción historial de la provincia y archipielago de Chiloé*. Santiago: Imprenta don Benito Cano.

Gosford, Robert. 2013. Fifty Years in the Desert – The Ethnobiological Life of Amadeo Rea. *Crikey*. Feb. 1. https://blogs.crikey.com.au/northern/2013/02/01/fifty-years-in-the-desert-the-ethnobiological-life-of-amadeo-rea/.

Hanisch, Walter. 1982. *Noticia breve y moderna del Archipielago de Chiloé, escrita por un misionero de aquellas islas en el año 1769 y 70. La Isla de Chiloé, capitana de rutas australes*. Santiago de Chile: Academia Superior de Ciencias Pedagógicas de Santiago.

Instituto Nacional de Estadísticas n.d.. Censos de Población Históricos. http://historico.ine.cl/canales/usuarios/censos_digitalizados.php.

Jiménez, Miguel. 2009. *Castro, Chiloé: Relatos de mi Niñez y Otros Escritos*. Santiago de Chile: Imprenta El Sur.

Klubock, Thomas Miller. 2014. *La Frontera: Forests and Ecological Conflict in Chile's Frontier Territory*. Durham: Duke University Press.

Laffoley, Daniel D'A., and J.M. Baxter. 2016. *Explaining Ocean Warming: Causes, scale, effects, and consequences*. Gland: IUCN World Commission on Protected Areas. https://portals.iucn.org/library/node/46254.

Miller, Jacob C. 2016. *Consumption, Dispersed: Techno-Malls and Embodied Assemblages at Chiloé Island, Chile*. University of Arizona. Unpublished dissertation.

Minnis, Paul E. 2016. Foreword. In *Ethnobiology of the Future*, edited by Gary Paul Nabhan. Tucson: University of Arizona Press.

Nabhan, Gary Paul, ed. 2016. *Ethnobiology for the Future: Linking Cultural and Ecological Diversity*. Tucson: University of Arizona Press.

Pitchon, Ana. 2015. Large-scale aquaculture and coastal resource-dependent communities: Tradition in transition on Chiloé Island, Chile. *Journal of Latin American and Caribbean Anthropology* 20 (2): 343–358.

Pfeiffer, Evelyn. 2016. Chile's Record Toxic Tides May Have Roots in Dirty Fish Farming. *National Geographic*. May 17.

Revista Salmonicultura. 2006. Cifras de la industria acuícola. *Revista Salmonicultura* 8 (78): 19.

Schurman, Rachel A. 2003. Fish and flexibility: Working in the New Chile. *NACLA Report on the Americas* 37 (1): 36–41.

———. 1996. Snails, southern hake, and sustainability: Neoliberalism and natural resource exports in Chile. *World Development* 24 (11): 1696–1709.

Sellner, Kevin G., Gregory J. Doucette, and Gary J. Kirkpatrick. 2003. Harmful algal blooms; causes, impacts, and detection. *Journal of Industrial Microbiology and Biotechnology* 30 (7): 383–406.

Soberanes, Rodrigo and Andrés Pérez. 2016. The Salmon Crisis on Chile's Chiloé Island. *Mongabay*. October 5. https://news.mongabay.com/2016/10/the-salmon-crisis-in-chiles-chiloe-island/.

United Nations. 2006. *A Case Study of the Salmon Industry in Chile*. New York/Geneva: United Nations Publications.

Urbina Burgos, Rodolfo. 2004. *Población Indígena, Encomienda, y Tributo en Chiloé, 1567–1813*. Valparaíso, Chile: Pontificia Universidad Católica de Valaparaíso.

Wallace, Tim. 2016. Oceans Are Absorbing Almost All of the Globe's Excess Heat. *The New York Times*. September 12.

Chapter 2
Local Knowledge, Local Networks, and Successful Cooperative Mussel Aquaculture on Chiloé

Ana Pitchon

As the world's wild fisheries are declining, and demand for seafood rising, aquaculture is taking the place of wild-capture fisheries by means of the controlled production of this important resource. While the continued supply of fish and shellfish was the original intention of aquaculture, the real driving force behind the industry's massive and rapid expansion is profit and centralized industrial production of food. Aquaculture is no different from agriculture and suffers the same consequences of industrialization. It is widely accepted that small-scale production is more socially and economically sustainable (Thu and Durrenberger 1998), but when this system is threatened by large-scale, exogenous global processes, local structures and identity are altered during the struggle for access to resources (Wells 1996). This can have severe consequences for local communities unaccustomed to change.

In Chile in particular, the marine ecosystem has proven to be a favorable environment for expansion in this field, at the regional, national, and international scale. Thus, areas in Chile where aquaculture is concentrated are in the process of sociocultural change as many traditional fishing families are making the transition from autonomous entities to participants in structured regimes. Because employment opportunities for these families are extremely limited, and their skills base so specific, subsumption into the dominant economic paradigm becomes inevitable. The impact of aquaculture in southern Chile, where the majority of the firms exist, has been significant on both an ecological and a social scale (Ohlson 1996; Caniggia 2000). While some social institutions remain intact despite this regime change (Barrett et al. 2002), changes on various scales have occurred in the coupled human-ecological system that is the Chiloé Archipelago.

In the following pages, I offer an overview of fishing and aquaculture in Chiloé. My discussion of fishing covers both artisanal finfish ventures and divers (*buzos*) who harvest wild shellfish. In describing aquaculture in Chile I reflect on large- and

A. Pitchon (✉)
Department of Anthropology, California State University Dominguez Hills, Carson, CA, USA
e-mail: apitchon@csudh.edu

A. Daughters, A. Pitchon (eds.), *Chiloé*, Ethnobiology,
https://doi.org/10.1007/978-3-319-91983-6_2

small-scale farming ventures involving salmon and shellfish. I also present a case study of the establishment of an aquaculture cooperative in a Chilote community, showing how small-scale, locally managed aquaculture firms of an endemic mollusk species can be not only an example of resilience but also a model for sustainable aquaculture, communities, and ecosystems.

Research Design

My research took place from 2004 to 2005 and again in 2017, and was carried out in two consecutive phases. Following the exploratory/explanatory design outlined by Johnson (1998), I used both qualitative and quantitative approaches to test the hypothesis that fishermen who were members of a mussel cooperative show higher indicators of resilience than fishermen who go to work for large-scale salmon firms. Though the types of data elicited in each phase were distinct, the complimentary contribution of qualitative and quantitative data provided for greater validity and depth.

I used two field sites for this research. I began with informal interviews over a 2-month period in the town of Chonchi, on Chiloé's big island. These interviews were conducted with anyone involved in the fishing and/or aquaculture industry, at all levels of the operational process and hierarchical structure. The purpose of these conversations was to introduce myself to the community, establish key informants, and understand the current situation from various points of view. Over time, I realized that in order to get a large enough sample, I would need to include the island of Lemuy, specifically the town of Puqueldón, as a part of the sampling frame. The island of Lemuy is within the jurisdiction of the Comuna of Chonchi, and also falls under the maritime responsibilities of the port of Chonchi Naval Authority. There is a strong connection between the two locations given that they share histories, families, and economies.

The first sample included individuals from both communities who were at various stages of transition from wild-capture fishermen to aquaculture employees. These included fishermen ($n = 53$), fishermen who had become aquaculture employees ($N = 119$), and those who were involved in both livelihoods ($N = 78$). They were also selected based on the amount of time they had been involved in those activities in order to reflect more accurately the temporal continuum. The total N was 277. Given the differing characteristics of each group, a standard systematic sampling design was not possible. Fishermen were selected using an exhaustive sampling procedure. Aquaculture employees were selected through a convenience sample. Those in the category of "both fishing and aquaculture" were identified via a question on the aquaculture questionnaire. The second sample group of mussel aquaculture cooperatives was also exhaustive ($N = 3$ cooperatives, 27 individual responses).

Study Area

The study area of Chonchi, Chiloé, and its environs has origins with the indigenous people known as the Huilliche who first occupied the area as small-scale subsistence agriculturalists and fishers more than a thousand years ago (see Chap. 1). Lifestyles of many families in and around the community today can be described as one of semi-subsistence, with fishing, foraging, and farming supplementing work in various market-related sectors. Cooperation is strong among Chonchi residents, which helps account for their continued overall well-being. However, many people do not have a positive view of the future, owing largely to declining natural resources and recurring threats to those resources, such as the red tide crisis of 2016 that paralyzed the economy of the archipelago (see Chap. 1). A comment by Don Jorge Chiguay in 2005 still reverberates today, particularly among older fishermen: "People before were more united, had more goodwill, and had enough food" compared with today. "People developed their land but without thinking of long-term development, since the future was less important than today. The problem is that with this vision, they also destroyed the land by clearing forests and over-working the land, leaving almost nothing to sustain these populations without help."[1]

Across the archipelago, many Chilotes refer to the days "before," which loosely means when they were younger, and before development began to occur. "Before we only had horses and boats to get around, there were hardly any people, maybe just a few houses, but the rest was only mountains," said Romelio Cárdenas in 2005.[2] Twelve years later, fisherman Hugo Mansilla of the island of Llingua had much the same to say: "We ate well, we had plenty of shellfish off our property, and the fishing was good. Today so much of the sea is contaminated, we can't do the same things."[3] Residents of Chonchi are still accustomed to counting on the availability of natural resources for basic survival, but they say that that availability is, like the rural Chilote way of life, becoming a thing of the past. "There used to be so much fish here, and every day we would go to the shore to *mariscar*, or collect shellfish, and it was covered with mussels, clams, snails, seagull eggs, and the fish could be caught without nets or line, just by grabbing them out of the water near the shore. We lived from working the land, grinding our wheat into flour, planting grains and vegetables – we always had these things, and we actually ate better then compared to now, and we didn't even have stoves, but used the *fogón* (smoke house)."[4]

Today, while some subsistence is possible, it is mainly a complementary activity as marine resources become ever more scarce and the market economy expands. People have had to look for resources away from home, creating a form of dependency on external resources and markets, either to sell their goods or to work. It is in this regard that rural Chilotes practice a strategy of semi-subsistence. They find

[1] Personal communication, Chonchi, 2005.

[2] Don Romelio Cárdenas, personal communication, Chonchi 2005.

[3] Hugo Mansilla, Llingua, December 2017 (interviewed by Anton Daughters).

[4] Don Raul Colivoro, personal communication, Chonchi 2005.

themselves in a state of transition between subsistence and market economies, often bemoaning the loss of traditional rural livelihoods as they adapt to the capitalist production process for survival. "People here aren't the same now. When someone needed something, there was always somebody to provide it, even if it meant going to town. Neighbors were very important and it was a beautiful thing. They were there to help, being happy with life."[5]

Local ecological knowledge in this region is profound, owing to the intense interaction with natural resources for survival over the generations (see Chap. 1). This knowledge—built up around the core livelihoods of fishing, shellfish gathering, small-scale agriculture, and the raising of livestock—functions as an integrated system of information on the interrelationships between animals and their natural environments. More important than the knowledge itself, however, is an ability to adapt to changing social and natural systems by developing new approaches for the continued use of resources (Warren and Rajasekaran 1993).

Fishing in Chile and Chiloé

Chile is a country with 78,563 km of coastline and 82% of its population living within 100 km of the coast (WRI Earthtrends 2005). It is second to Peru for wild-capture fisheries in South America, producing half the total yield, but with indications of a decline in the industry (FAO 2015). In its 2017 Review of Fisheries, OECD registered a downtick in exports of wild-capture fish in 2016, at least partly the consequence of the red tide outbreak of that year; since 2006, however, total exports have remained relatively steady, just under US$2 billion a year (OECD 2017). As a point of comparison, aquaculture production in Chile has undergone rapid and significant growth, increasing by 1360% since its inception in 1980 (World Resources Institute 2015).

The majority of the fisheries in Chile are small-scale artisanal. Of the more than 50,800 people employed in both the industrial and artisanal fishing sectors, about 45,700 of them are artisanal. Their catch comes to 20% of total exports of fish products in any given year (a figure that also takes into account aquaculture exports) (FAO 2015). However, sharp declines in finfish populations have led to a reduction in the artisanal fishing sector in recent months, forcing some islanders to seek employment elsewhere. When and if wild stocks rebound remains to be seen.

The national trend of decline was already in evidence in Chonchi in 2004. The fishermen there provided the same story each time I spoke with them. According to one multi-generational fisherman,

> No one wants to be a fisherman anymore. It is very sacrificial work. I certainly don't want my son to be a fisherman. Every year it gets worse, the sea is getting worse, and our fish are disappearing. Today, I got 300 kilos of sea urchin. Five years

[5] Don Raul Colivoro, personal communication, Chonchi 2005.

> ago we were getting 800 kilos a day. The price has only gone up ten pesos in ten years, from 190 to 200 pesos per kilo. We have to go farther away, sometimes travel all night, just to get the same amount of product that we used to, and that means we work more hours. It is more dangerous, because we have to go so far, where the sea is more dangerous, and we have to dive deeper there too. I'm going to buy a cow.[6]

There are approximately 24,000 artisanal fishermen in Chile's Lakes Region, with an estimated 10,000 in Chiloé (INE 2008). These comprise a significant percentage of the region's total population. Historically, their overall impact on wild stocks was largely negligible—until the early 2000s, their fishing boats had no motors and were only about 2 or 3 m long. Today, artisanal boats range in size from 3 to 18 m[7] and have outboard engines. Though they have the capacity to venture farther offshore than before, they are legally required to remain within five miles of the shoreline.

Given the fluctuations in wild stocks, these fishermen often find themselves at the greatest economic disadvantage. In Chonchi, the number of artisanal fishing boats has dwindled over the years. Those fishermen who are still active do so mostly as a supplemental activity; few of them work exclusively in commercial fisheries today. Those few holdouts are often helped by their wives who are employed primarily in the service sector. With one or both spouses working for wages full time, fishing is carried out on weekends, early mornings or evening.

Altogether, the conditions faced by artisanal fishers in Chiloé are challenging and present a feedback loop from which most can never escape. Even though fishermen cooperate and help one another, this is not enough when a motor dies and there is no money to replace it, when red tide closes the mollusk fishery for months at a time, or when farmed products are preferred to wild catches for issues such as uniformity and traceability.

Buzos (Divers)

There is another category of fisherman, however, that is still managing to scrape out a living in this line of work, though their days appear numbered. These are the *buzos* or divers, who have more complex technology and larger boats. They are also classified as artisanal fishermen, with boat size restrictions and different catch limits than those of the industrial fleet. They operate near-shore, though as stocks have been depleted, they have had to venture farther and farther out—and farther from their homes—to find their catch. Their boats are not open dingy type vessels like many of the smaller finfish fishing boats, but rather have an open end and an enclosed bridge and below-deck galley and bunk room. These fishermen dive for their catch,

[6] Personal communication, Chonchi 2004

[7] Sernapesca defines an artisanal boat as one that is less than 18 meters in length.

Image 2.1 Don Pedro Cardenas, a *buzo* in Chonchi, in 2006. (Photo credit: Ana Pitchon)

rather than use hooks, lines, and nets. They wear inch-thick neoprene wet suits, attach a 40 m hose to a mask, and typically have a first mate pump and watch the propane generated oxygen tank fixed on the back of the vessel. This practice of diving for marine resources goes back generations, and divers historically used sea lion fat for insulation. Divers are restricted to a depth of 20 m, although many dive deeper for the same reason that they increase their distance from home (Image 2.1).

The *buzo's* catch consists of mussels, clams, sea urchins, crabs, seaweed, and some conger eel. There are seasonal regulations and catch limits for these species, though I observed that most species were a year-round commodity if there was a buyer. Though their season was short, sea urchins were the most lucrative, with an export market primarily in Japan. Mussel and clam ventures were periodically in distress, not because of limited catch size but because competition with aquaculture products made it more difficult for divers to find a buyer. Increasingly frequent bouts of disease, such as red tide and other bacterial contaminants (most notably *Vibrio parahaemolyticus*) also placed stress on these species. Farmed mollusks are not immune to such pathogens, but with more frequent water and product testing and traceability standards, aquaculture operations are more likely to encounter success in the national and international markets.

Fishers in Chonchi

Artisanal fishermen in Chonchi, including buzos and finfish fishers, are at a particular disadvantage due to the geography of their town. Its location has proven favorable for salmon and mussel farms due to efficient currents and water depth, as well as shelter from strong winds, all of which lend to success in aquaculture. There is only one port in the town, and it is shared by artisanal fishermen and the larger vessels belonging to the farms that ferry smolt, feed, salmon, and people to and from the farms.

At a meeting held by the mayor with artisanal fishermen in March 2005, it was mentioned that the port was to be privatized, with ownership given to the farms for docking rights. The fishermen felt this was a final blow to their industry, but with no established sea tenure, nor significant economic contribution to the town, there is almost nothing they can do to protect themselves against this redistribution and privatization of common property.

There is much contention between fishermen and the aquaculture firms all over the archipelago. The fishermen are extremely suspicious of the government and the corporations, and usually for good reason. Most are very much aware of some detail of inequality in the system. For example, the predecessor to the current *Subsecretario de Pesca* (Secretary of Fisheries – the highest national office responsible for overseeing the regulation of fishing activities) was for many years president of the Association of Salmon and Trout Producers of Chile, a pro-industry lobby group. In fact, *Subsecretario* Daniel Albarrán was pressured to resign from his national appointment in August 2001 for allegedly receiving bribes from the salmon exporting company Aguas Claras (Aquanoticias 2002; Ecoceanos 2002). Similar scandals have plagued the government in the intervening years.

My fieldwork in 2005 showed that acts of defiance against the aquaculture firms were not uncommon. Fishermen routinely broke salmon nets, not as an act of vandalism as much as to temporarily increase the quantity of high market value species. Open-ocean salmon fishing was—and continues to be— prohibited, though when asked, most fishermen could not give me a straight answer about its legality. Even though research shows that allowing a wild-capture salmon fishery could potentially solve many environmental problems associated with the escapees (Soto et al. 2001), it has yet to be legalized.

Despite their adversarial view of salmon farms, some people I spoke with in Chonchi also saw them in a positive light. "At least they are here, offering employment. Before, the men used to just stand around on the corner, doing nothing. Now, even though we are still poor, at least we have work."[8]

In 2005, I set out to test the hypothesis that fishermen who become employees in aquaculture firms tend to score lower in certain indicators of resilience (i.e., job satisfaction, perceived well-being and environmental values) than those who remain fishermen, despite the extreme difficulty of making a living. As discussed above, I chose three groups to represent this transition: fishermen, aquaculture employees, and people involved in both activities. The hypothesis was based on the assumption that autonomy is the principle characteristic of satisfaction among fishermen (Acheson 1981; Pollnac 1997; Orbach 1997), and that autonomy tends to be lost when fishermen become wage laborers.

[8] Personal Communication, Chonchi 2004.

The results from analysis of this research indicated that the closer along the trajectory one got to being an aquaculture employee, and the farther away from being a fisherman, the lower the overall scores across all three indicators. Although fishermen who worked at the salmon farms received a steady paycheck, they had lower levels of job satisfaction and overall well-being. However, those involved in the implementation and management of their own aquaculture cooperative (discussed in more detail below) showed the highest scores of all three.

Aquaculture in Chile and Chiloé

During the past two decades, aquaculture has developed significantly in Chile, though it has been an active industry in the region since the early 1960s (see Chap. 1). Aquaculture is generally perceived to be a profitable activity, economically and socially, and has thus established itself as an industry destined for expansion. But it has triggered dramatic social change since its inception and rapid growth, most notably in traditional fishing communities along the southern coast of Chile and the Archipelago of Chiloé. Therefore, the future of aquaculture in the region will be highly dependent upon the resilience of the industry's economic and social institutions.

Aquaculture has changed the seascape of the Chiloé archipelago as much as agriculture has on land, though at a much faster rate. It appears, when looking across the bays from land, as though plots of farmland have slid into the water, as the squared-off individual concessions give the appearance of crops in the water. Navigation of larger vessels has become more difficult since the sea is now cordoned off into maritime concessions. The shoreline is littered with plastic and foam pieces that have been broken off of the floats that suspend mussel chains. The salmon companies employ armed guards to deter theft and net breakage, while the smaller farms have their own means of theft and vandalism deterrence, usually a 24-h watchman. Unguarded, a concession is quite likely to be robbed, not so much out of desperation for the product, but more out of spite and a feeling of vengeance by local residents. Granted, if there is no threat of an armed guard, taking a chain full of mussels is a lot easier than spending several hours underwater looking for them (Image 2.2).

There are some aquaculture activities in the world that are destined to be an aid in socioeconomic development among the rural poor. In Chile, however, the export industry is relatively young, its large-scale operations driven largely by economic concerns, not social ones. Unfortunately, the primary aquaculture product in Chile, salmon, is an elite product destined for only a small percentage of the world's population. This is problematic for several reasons. First, what should be a widely available protein source is now an elite consumer item. These are salmonids in Chile, and shrimp in Ecuador, Thailand, and India. Much of the new research and development worldwide is still focused on these high-end consumer products, mostly exported to Europe, the United States, and Japan. (Mussel aquaculture in Chile, it should be

Image 2.2 View of seascape and mainland from Quellón. (Photo credit: Ana Pitchon)

noted, is an exception to this rule.) Second, the technological overhead for farming many species, especially carnivorous fish like salmon, is so immense that only corporations can afford to run their production. In the case of Chile, these corporations are mostly foreign-owned. Third, the corporations rely on local labor and return little to the local economy. Wages are low, and workers are hired and fired with frequency. In short, they deplete the local environment and limit access to resources, while channeling capital away from its source.

The majority of the aquaculture firms on Chiloé are salmon producers (*Oncorhynchus mykiss, Oncorhynchus kisutch*, and *Salmo salar)*, followed by mollusks (three species of Mytilids - *Mytilus chilensis, Choromytilus chorus*, and *Aulacomya ater*), oysters (*Tiostrea chilensis* and *Crassostrea gigas*), scallops (*Argopecten purpuratus*), and seaweeds. However, mussels have recently surpassed salmon in terms of percentage of the total aquaculture harvest in the country, at 43.5% compared with salmon at 42.5% (FAO 2015).

In 2016, Chile's Subsecretaría de Pesca (Subpesca) counted more than 600 shellfish producers in Chile's Lake Region, more than 80% of them classified as small- or medium-scale operations (Aqua 2017). These ventures employ thousands of islanders, and in 2017 harvested more than 23,000 tons of mollusks (INE 2018). Employment conditions are not ideal, but as many islanders have told me over the years, at least the jobs are there. The challenge, they stress, is keeping the growing industry from disrupting communities and the fundamental beneficial attributes of rural Chilote culture.

The most widely farmed species in the shellfish category are mussels, followed by oysters and scallops. Mussel farming is one of the oldest aquaculture activities in Chile, beginning experimentally in 1943 in the southern sector of Chiloé's big island. The government established the first mollusk farming operations on rafts and long lines in other areas of the Lakes Region in 1967, using French and Spanish technology. This effort was the first state operated aquaculture venture, and was run by the Institute of Fisheries Promotion (IFOP) and the Division of Fishing and Hunting, under the department of agriculture (Servicio Agrícola y Ganadero, or SAG) (Aquanoticias 2002).

Until the early 1980s, this original system changed little, and was managed largely on a small-scale basis. Despite the participation of the Universidad Austral de Chile—with the goal of drawing the private sector into this activity— advances were still marginal.

In the 1980s, larger firms started to drive production and distribution of salmon and shellfish. Around this time that the native species of *Mytilus chilensis* (*el chorito*) was starting to be farmed, due to increasing demand and declining wild stocks. Since then, choritos have accounted for the majority of farmed mussel production in Chile. Unlike the salmon farms, mussel aquaculture is still dispersed, meaning that in the majority of the cases, seed farms, species cultivation sites, processing plants, and exporters are separate entities, neither functioning as a whole nor as a single corporation.

Foreign interest in Chilean aquaculture has grown significantly in the past two decades. Spanish companies first entered the market during the latter part of the twentieth century with comprehensive processing plants for mussels, establishing an important industrial purchasing power. This shift from artisanal, local practice to industrial-scale, international production was massive and abrupt, bringing with it an explosion of mussel aquaculture centers. Spain is still the primary market for the Chilean mussel industry. In 2004, exports to Spain were worth US$32 million, double the amount from the year before (El Llanquihue 2004). By 2017, that figure had climbed to more than $50 million (FIS 2017). Worldwide, farmed mussels account for more than 70% of total mussels sold over the past decade, with Chile leading all other countries in production (FAO 2016). The growth of mussel farms in the region shows no signs of slowing down as Chile remains the top exporter of the species.

As a product that is easily farmed, there is much worldwide competition in the production and marketing of mussels. Despite high levels of production, the Chilean mussel industry faces some strong obstacles on the international market, such as shipping distances, and increasingly frequent outbreaks of red tide (see Chap. 1). However, the most common species farmed (*Mytilus chilensis*) is favored in some markets, keeping it relatively strong. As with salmon farming, this industry is concentrated in the Lakes Region and Aysén due to water quality and geographical conditions.

The mussel industry's burgeoning popularity has inspired more cooperation than usual out of the need to deal with the high costs of monitoring and marketing. The Association of Mollusk Farmers of Chile (AmiChile) is an excellent example of this. The association consists of firms of all sizes, but non-corporate and mostly family owned. It has been in existence since 1991, with the initial objective of establishing a group to create a nexus between its members and the government. It began with 11 producers, all family firms that cooperated for a voice in legislation and to facilitate marketing. By 2016, it had grown to include 59 producers. As the industrial sector discovered that mussel farming was a simple and lucrative resource, the venture became more competitive for these small, family-run farms, but the strong demand for mussels worldwide has allowed the association to continue to grow.

Cooperation, as exemplified by AmiChile, is the best way for these firms to compete in the growing market. However, small-scale producers have not yet reached the level of coordination found with industrial firms, where all services are housed together, from seed to processing to export. Therefore, they are forced to go through middle-men for most services, which in turn raise the costs of production and the prices they charge.

For the small firms I researched in 2005, the production costs amounted to 60 pesos a kilo; their net gain per kilo was 40 pesos, yielding them a profit of 20 pesos, or approximately $0.37 per kilo of final product. Clearly, cooperation would have given them an advantage. But based on conversations with members of the association and fishermen, personality differences deterred them from cooperating.

History shows that fishermen are not actually too independent to self-organize, as the folk model would indicate. In the USA in the 1930s, fishermen unions were redefined as firms so that the government could fix prices (Durrenberger 1996). Today, survival of the family firms depends more and more on the ability to cooperate, and the attitudes I encountered in 2005 are slowly changing in an effort to remain competitive. The benefits of being a member of the association are much like those of a cooperative, though the members do not see it as such, since they do not share production duties cooperatively. Like a cooperative, however, by joining forces at this administrative level, they are given institutional power and a voice that they otherwise would not have.

The Mussel Cooperatives

The cooperatives that I investigated in 2005 were established with initial financial and technological aid from the Fundación Chinquihue, a nonprofit organization subsidized by the government of the Lakes Region of Chile. The objective of this program was to provide an alternative livelihood for artisanal fishermen in Chiloé who were experiencing difficulty in maintaining their traditional craft. By encouraging them to adopt mussel aquaculture and other bivalve species of high commercial value, artisanal fishers could establish greater economic stability.

The initiative was based on a larger effort by the International Center for Development Research (CIID) and the Marine Research Center and Institute of Meat Science and Technology of the University Austral de Chile, a research branch aimed at providing a development scheme for rural coastal communities on the big island of Chiloé engaged in mariculture. Of note was the fact that the program made a significant effort to understand the cultural and psychocultural characteristics of these rural coastal communities. Special attention was paid to their responses to distinct alternatives to production.

The Fundación Chinquihue focused exclusively on mussel (*Mytilus chilensis*) cultivation in Chiloé, facilitating production efforts by using a native species. The foundation, from its headquarters near Puerto Montt on the mainland, grew its own seed and provided on-site processing. It also offered free services to the cooperatives formed under its initiative. To find willing communities, the foundation did some outreach among fishermen's cooperatives. It provided financial and technological support, as well as education on the process. However the fishermen were ultimately responsible for organizing themselves into a unit and securing the various permits.

Fishermen were also responsible for supplying the funds to lease the water space for their operations. This was where some communities ran into trouble. While all of the communities I questioned (N = 7) were agreeable to this effort, the individual personality characteristics of each group affected the viability of the project. For example, in Chonchi, the fishermen were self-proclaimed individualists, and even though they were excited and interested in the idea, they were simply unable to organize around this effort despite repeated attempts on my behalf. Therefore, the effort was established on the neighboring island of Lemuy in the town of Puqueldón, where the fishermen were more open to cooperating.

Fishermen who were members of the mussel aquaculture cooperatives registered significantly higher scores across all three indicators (PWB – 31.42; EV – 78.35; JS – 64.87[9]). Even though these fishermen had structurally changed their occupation, basic elements of their foundations persisted, such as autonomy, use of preexisting knowledge, and the maintenance of social networks. These characteristics, I argued, could provide a foundation for resilience in the face of the significant changes taking place across the island with regard to ecosystem services and access.

Altogether, the study found that the development program outlined above did change the structure of these coastal subsistence communities, though in ways that were positive for social sustainability. Having increased economic stability actually aided fishermen in maintaining traditional subsistence agriculture by allowing for the purchase of animals, home improvement products, and fishing and diving materials. In this regard, the standard of living of these coastal communities improved. The adoption of mussel cultivation, moreover, had a positive impact on the ecosystem. Fishermen applied a management strategy that protected the resource by repopulating the species for future harvesting. The practice also strengthened a conservation ethic among fishermen by giving them a greater understanding of the issues involved in the life-cycle and ecosystemic participation of the species.

[9] Data were analyzed in SPSS. One-way ANOVA and the Dunnett's T3 post-hoc test were used to produce results that looked for and compared significance across the groups and subgroups, which were the social-ecological indicators (dependent variables) and occupation (independent variables).

Is Aquaculture Sustainable?

The future of aquaculture depends on cooperation between stakeholders, including regulatory agencies, universities, scientists, and fishermen. Improving technology and research and development into species diversification can promote a sustainable future of marine resource consumption and may benefit a wider range of consumers. The diverse ecological, sociocultural, and political interests involved make this a challenge, though one that has the potential to be met. Global cooperation is paramount for the diffusion of successful information and technology to establish and maintain sustainable practices. Under the right conditions, aquaculture has the potential to meet demand for this important resource, while at the same time establishing socioecological improvements that will benefit these systems on a global scale.

Commercialization of common pool resources has long been known to have adverse effects on local peoples and often creates dependency on external structures or institutions (Gunder-Frank 1967; Greenberg 1998). Much of the literature on commercialization of resources focuses on land-based resources, due to the fact that agriculture has been around significantly longer than aquaculture. However, the two practices are strikingly similar. In both agriculture and aquaculture, local resources are often used to produce feed and create space, the environment is contaminated and often destroyed, and the local communities once dependent on these resources suffer economically from an inability to sustain their livelihood in historically traditional ways. Similarities are even found with respect to the alteration of land/seascape through the development and allocation of plots.

The case of Chile is somewhat unique in that aquaculture was not initially introduced as a development tool, unlike countries such as India, Thailand, and Ecuador. Aquaculture came to Chile when the country's natural resource base and amenable marine ecosystems were discovered by large-scale international producers. International corporations combined forces with the Chilean government to establish large-scale aquaculture in the region, primarily destined for the export market. Aquaculture in Chile was never meant to ease the food production burden, since there never has been a shortage that demanded such schemes.

However, the differences end there. Commodification on this scale had led to increased poverty and increased dependence on external institutions as a result of loss of marine access. Aquaculture continues to be considered a part of fisheries, despite being an independent sector, more akin to agriculture (Pillay 1994), due only to its association with marine resources. This is why many agencies charged with the management of such species have decided that this would be an appropriate transition for people dependent on wild-capture fisheries who are faced with an economic and social crisis due to the decline and/or closure of their fisheries. In the USA, the National Oceanic and Atmospheric Administration (NOAA), the agency that houses NOAA Fisheries and regulates federal fisheries, has suggested this strategy (NOAA 2005), as well as other regional governments in Chile facing similar circumstances (El Llanquihue 2004). While superficially this appears to be an excellent solution in the face of degraded wild-capture fisheries, there are inherent and fundamental differences between the two practices that may make this far from ideal.

Keeping people connected to the sea, where their knowledge and skills base and a feeling of competency and ownership can continue, is of paramount importance (Pollnac and Poggie 1988; Diener 1997). The direction of such plans, then, should be carefully considered. When fishermen become employees of large firms, such as the situation in Chile, much of this is in fact lost. Perhaps a better solution is to provide them with a level of autonomy comparable to what they have as fishermen by forming small-scale aquaculture operations where the same degree of ownership and knowledge is maintained.

Future Direction

There is still a lack of agreement as to whether aquaculture is a viable solution to the disappearance of wild-capture fisheries, and whether it can serve as an income alternative. The persistent idea that fishermen are too culturally dissimilar from the practice of controlled cultivation for it to be a viable solution is the primary reason behind such thinking (NRC 1988; Pollnac and Poggie 1982; Pollnac and Weeks 1992). There is new evidence of success in this strategy, which may be due in part to an in-depth cultural understanding and compatibility, and perhaps most importantly, a need (Pitchon 2011; Outeiro and Villasante 2013; D'Anna and Murray 2015). The success in the case of Chiloé is due to the foundation's success in following these guidelines, but also in their insistence on a "hands-off" approach that allows for ownership at a level beyond that of participation. This is not to say that participation is not a critical component, and the approach taken by the foundation may not be right for every circumstance, once again stressing the importance of sociocultural compatibility.

According to Pomeroy (1992), these initiatives should be implemented as a complimentary activity to fishing. The difficulty with this assertion is that in many instances fishing is no longer an option due to closed and/or nonexistent fisheries. One relevant point expressed by Pomeroy is the desirability of agricultural expertise or at minimum, understanding, that would aid groups of fishermen in making the transition since aquaculture is more akin to agriculture than to fishing (1992). I have expressed this as an important component as to why the initiative has been successful in Chiloé, given the historic dual-activity lifestyle of working both the land and the sea. Using agricultural knowledge and techniques that Chilotes have with water-based activities has eased the transition and has allowed for some success.

Fisheries worldwide are in a dire position, yet the demand for seafood continues to rise, making aquaculture a logical step in supplying this resource. Clearly, aquaculture is becoming an increasingly important economic activity and food production strategy throughout the world, in the North and South, and in temperate and tropical regions.

Intensive, large-scale operations, especially those that farm nonnative species, can be detrimental to the human ecosystem that they are in. At the same time, international development agencies and state and local governments herald aquaculture as a means of economic development, resource diversification, and food security, and some scientists argue that it can be a strategy for taking the pressure off of wild fish stocks in order for them to recover.

This all may be true, but special attention needs to be paid to planning and conceptualization of these strategies at the community level, and that a long-term vision should be in place that ensures a resilient and sustainable operation. Social-ecological systems can be sustained in spite of perturbations such as the one discussed here by the introduction of aquaculture, but acknowledging the importance of resilience and its characteristics is critical to success. Though my research is in Chile, I use standard methodologies for empirical measurement to allow for replication and also comparison with other sites to allow for generalization. This then can have important implications for maritime and coastal policy both in Chile and elsewhere where aquaculture or other disturbances are supplanting small-scale fisheries and altering communities. Aquaculture cooperatives may not be the panacea for all fishing communities under pressure to find alternatives, but I have demonstrated that in this case it is a viable and sustainable substitution for maintaining a sense of basic cultural and community structures in Chiloé.

While we see dramatic changes when cultures transition from terrestrial foraging to horticultural activities, the dynamics of groups reliant on marine resources differ. We can not necessarily expect to see similar transformations. Future research should explore the resilient nature of such communities in the face of potentially dramatic shifts in economic production and address the factors that maintain community structure. Within this line of thought, some questions arise, such as, under what conditions are these transitions sustainable and result in improvement of livelihood? How does cooperation change in this transition? Are cooperative firms competitive with capitalist firms? And also, how does community structure change or how it is maintained? In the case of Chiloé, a social network analysis would be beneficial, as would a more in-depth analysis of the changes in state variables and scales involved in the transformation.

Ultimately, it is up to the individuals within communities in a state of change to be amenable to and promote resilient strategies for sustainability. The fishermen of Chiloé are working toward this goal, and many have experienced success (see Chap. 3). Given the right parameters, it is possible to transform without being harmful or disruptive to the primary ecosystem of which they are a part. This was possible because of the flexibility of management and the willingness to be adaptive, both of which are critical components of successful policy. Despite criticism of such approaches because of time and economic constraints, this case has shown that neither need be impediments to resilience, and that with foresight and some basic variables in place, resilience and sustainability are possible for coastal communities facing change.

References

Acheson, James. 1981. Anthropology of fishing. *Annual Review of Anthropology* 10: 275–316.
Aqua. 2017. Moluscos y Algas en 2016: Los otros cultivos que mueven la acuicultura nacional. May 17. http://www.aqua.cl/reportajes/moluscosalgas-2016-los-otros-cultivos-mueven-la-acuicultura-nacional/.
Aquanoticias. 2002. Producción. March. Santiago, Chile.
Barrett, Gene, Mauricio Caniggia, and Lorna Read. 2002. There are more vets than doctors in Chiloé: Social and community impact of the globalization of aquaculture in Chile. *World Development* 30 (11): 1951–1965.
Caniggia, Mauricio. 2000. Evaluación ex post facto del programa de desarrollo de la maricultura y procesamiento de moluscos en comunidades rurales costeras de Chiloé-Chile. Tesis para optar al grado de Magister en Desarrollo Rural, Universidad Austral de Chile.
D'Anna, Linda, and Grant Murray. 2015. Perceptions of shellfish aquaculture in British Columbia and implications for well-being in marine socialecological systems. *Ecology and Society* 20 (1):57.
Diener, Ed. et al. 1997 Subjective well being. Indian *Journal of Clinical Psychology*.
Durrenberger, Paul E. 1996. *Gulf Coast soundings: People and policy in the Mississippi shrimp industry*. Lawrence: University Press of Kansas.
El Llanquihue. 2004. Domingo, 21 de Noviembre.
Ecoceanos News. 2002. Bajo graves acusaciones de corrupcion renuncio el subsecretario de pesca. Santiago de Chile. August 30.
FAO. 2015–2016. Food and Agriculture Division of the United Nations. Fisheries Sector. http://www.fao.org/fi/fcp/en/CHL/profile.htm; http://www.fao.org/figis.
FIS. 2017. Chilean mussel exports grow in volume and value. Fish Information Services. *December*: 22 http://www.fis.com/fis/worldnews/worldnews.asp?monthyear=12-2017&day=22&id=95346&l=e&country=&special=&ndb=1&df=1.
Greenberg, James B. 1998. The tragedy of commoditization: Political ecology of the Colorado River Delta's destruction. *Research in Economic Anthropology* 19: 133–152.
Gunder-Frank, Andre. 1967. *Capitalism and underdevelopment in Latin America*. New York: Monthly Review Press.
Instituto Nacional de Estadísticas (INE). 2008 and 2018. Boletín de Pesca, Región de los Lagos. Santiago: Chile. http://www.aqua.cl/wpcontent/uploads/sites/3/2018/03/Bolet%C3%ADn_Los_Lagos_Diciembre_2017.pdf.
Johnson, Jeffrey. 1998. Research design and research strategies. In *The handbook of methods in cultural anthropology*, ed. H.R. Bernard. Walnut Creek, CA: Altamira Press.
NOAA. 2005. Aquaculture. www.nmfs.noaa.gov/mediacenter/aquaculture. Accessed June 2005.
National Geophysical Data Center (NGDC), NOAA. 2005. http://www.ngdc.noaa.gov/seg/hazard/slideset/45/45_slides.shtml. Accessed June 2005.
National Research Council (NRC). 1988. *Fishing technologies for developing countries*. Washington, D.C.: National Academy Press.
OECD. 2017. *OECD review of fisheries: Policies and summary statistics 2017*. Paris: OECD Publishing. https://doi.org/10.1787/rev_fish_stat_en-2017-en.
Ohlson, Willliam. 1996. Los impactos de las salmoneras en Chiloé. Unpublished Document. Archivos de Chiloé. Castro, Chile.
Orbach, Michael. 1997. *Hunters, seamen, and entrepeneurs: The tuna seinermen of San Diego*. Berkeley: University of California Press.
Outeiro, Luis, and Sebastian Villasante. 2013. Linking salmon aquaculture synergies and trade-offs on ecosystem services to human wellbeing constituents. *Ambio* 42 (8): 1022–1036.
Pillay, T.V.R. 1994. *Aquaculture development: Progress and prospects*. New York: John Wiley and Sons, Inc..
Pitchon, Ana. 2011. Sea hunters or sea farmers: Transitions in Chilean fisheries. *Human Organization* 70 (2): 200–209.

Pollnac, Richard B. 1997. Social and cultural characteristics of fishing peoples. *Marine Behavioral Physiology* 14: 23–39.

Pollnac, Richard B., and John Poggie. 1982. Sociocultural aspects of implementing aquaculture systems in marine fishing communities. In *Aquaculture development in less developed countries*, ed. L.J. Smith and S. Peterson. Boulder: Westview Press.

———. 1988. The structure of job satisfaction among New England fishermen and its application to fisheries management policy. *American Anthropologist* 90 (4): 888–901.

Pollnac, Richard B., and Pricilla Weeks, eds. 1992. *Coastal aquaculture in developing countries: Problems and perspectives*. Kingston, R.I: University of Rhode Island, ICMRD.

Pomeroy, Robert S. 1992. Aquaculture development: An alternative for small-scale fisherfolk in developing countries. In *Coastal aquaculture in developing countries: Problems and perspectives*, ed. Richard B. Pollnac and Pricilla Weeks. Kingston, R.I: University of Rhode Island, ICMRD.

SERNAP. 2005. Servicio Nacional de Pesca. http://www.sernapesca.cl/. Accessed June 2005.

Soto, Doris, Fernando Jara, and Carlos Moreno. 2001. Escaped salmon in the inner seas, southern Chile: Facing ecological and social conflicts. *Ecological Applications* 11 (6): 1750–1762.

Thu, Kendall M., and E. Paul Durrenberger, eds. 1998. *Pigs, profits, and rural communities*. Albany: State University of New York Press.

Undercurrent News. 2015. Chile mussel exports grow in volume and value. https://www.undercurrentnews.com/2015/07/10/chile-mussel-exports-grow-in-volume-and-value/. Accessed March 2018.

Warren, D.M., and B. Rajasekaran. 1993. Putting local knowledge to good use. *International Agricultural Development* 13 (4): 8–10.

Wells, Miriam J. 1996. *Strawberry fields: Politics, class, and work in California agriculture*. Ithaca, NY: Cornell University Press.

World Resources Institute. 2015. Earth Trends Data Tables. http://earthtrends.wri.org/datatables. Accessed August 2005 and January 2018.

Chapter 3
Livelihood Diversification as a Form of Resilience? An Ethnographic Account of Artisanal Fishers in Chile's Lakes Region

Sarah A. Ebel

For many artisanal fishers worldwide, the act of fishing goes beyond the need for income and is essential to their ways of life, their ability to manage risk, and their likelihood of reducing poverty (Marschke and Berkes 2006; Peterson 2014; Urquhart and Acott 2013). Artisanal fishers—individuals who use small boats with little technology[1] to fish areas near shore—face many challenges and threats to their way of life, including declining fish populations, poor resource governance, and climatic and ecological change (Beddington et al. 2007). To manage this risk while still maintaining their social and cultural ties to fishing, many fishers diversify their livelihood strategies (Allison and Ellis 2001; Ellis 2000). Livelihood diversification is "the process by which households construct an increasingly diverse portfolio of activities and assets in order to survive and to improve their standard of living" (Ellis 2000:14). The diversification of livelihood strategies offers fishers ways to spread the risk of fishing and other opportunities for sustenance over multiple sources. This can sometimes increase their resilience—an individual's or socioecological system's ability to cope with, and adjust to, disturbances in the marine economic, political, or ecological systems (Allison and Ellis 2001; Goulden et al. 2013; Holling and Gunderson 2002; Walker et al. 2004).

This chapter presents an ethnographic and empirical analysis of the livelihoods of artisanal fishers in Chiloé and the broader Lakes Region after the crisis of 2016.[2] The crisis dealt a significant blow to artisanal fishers' economic and social well-being, bringing to the surface preexisting tensions between fishers, aquaculture companies, and the government (Barrett et al. 2002; Barton and Fløysand 2010). I emphasize fishers' perceptions of social and ecological change, and the importance, but limitations, of a strategy of diversified livelihoods. The research is based on

[1] Artisanal fishing boats typically range in size from 9 to 30 feet.

[2] For more on the crisis, see Chap. 1.

S. A. Ebel (✉)
Department of Anthropology, University of Maine, Orono, ME, USA
e-mail: sarah.ebel@maine.edu

A. Daughters, A. Pitchon (eds.), *Chiloé*, Ethnobiology,
https://doi.org/10.1007/978-3-319-91983-6_3

13 h of recorded interview data, 3 months of field work in the Lakes Region as a participant observer in 2016, and 5 months of experience as a volunteer for Chile's National Confederation of Artisanal Fishers in 2010. Through the fishers' own words, I show the importance of fishing to their identities and their social, cultural, and economic well-being, and I stress the risks of relying on livelihood diversification as a form of social and economic resilience in a dynamic, globalized world.

Artisanal Fisheries in Chiloé and the Broader Lakes Region

Chile is home to over 90,000 artisanal fishers who live in diverse social and ecological systems from the Atacama Desert in Chile's north to the fjords and windswept fields of Patagonia in the south. Many are economically dependent upon marine resources and have strong social and cultural ties to fishing. Every year since 2008, artisanal fishers have harvested more fisheries product than Chile's industrial fishing fleet, making a substantial contribution to Chile's economy (Moreno and Revenga 2014).

On the Archipelago of Chiloé and the broader Lakes Region, approximately 24,000 fishers are members of community-based fishing unions, formed under Chile's fisheries and aquaculture law, *La Ley de Pesca y Acuicultura* (INE 2008). Passed in 1991, and revamped in 2012, the law established geographical areas where a union's members are legally allowed to fish. The majority of fishers within these designated areas use diving as a harvesting method to extract the economically and culturally important *loco* (abalone *Concholepas conchoelpas*), sea urchin, crab, and octopus, among other species. Along the shoreline, others gather seaweed. A diver uses a mask and a *chinguillo*, which is a bag that attaches to his or her waist where the diver places the harvest. While under water, the diver breathes through a tube called a *hooka*. The *hooka* runs up through the water to the boat, where it is attached to an air compressor. The fishers dive off small open-boats, usually made of wood and powered by a small outboard motor, or are human-powered by rowing. One or two individuals accompany the diver, and stay in the boat to maintain boat control, regulate the air compressor for the diver, and sort the harvested product. These individuals are known as tenders. The boat and its occupants bring back their harvests to their fishing unions and sell their product to local, regional, and national distributors.

The Lakes Region is the most fisheries-dependent region in Chile (Moreno and Revenga 2014), and a majority of the communities in the region are rural and poor (Latta and Aguyayo 2012). Many of these rural households have adopted diversified livelihood strategies rooted in longstanding relationships with the natural environment. Individuals work as artisanal fishers during the *loco*, sea urchin, crab, and clam seasons and simultaneously engage in subsistence agriculture to feed their families and their neighbors (Latta and Aguyayo 2012).

Artisanal fishers launch their fishing vessels from ports in coves, called *caletas*. The *caletas* discussed in this chapter are located in the mainland communities of

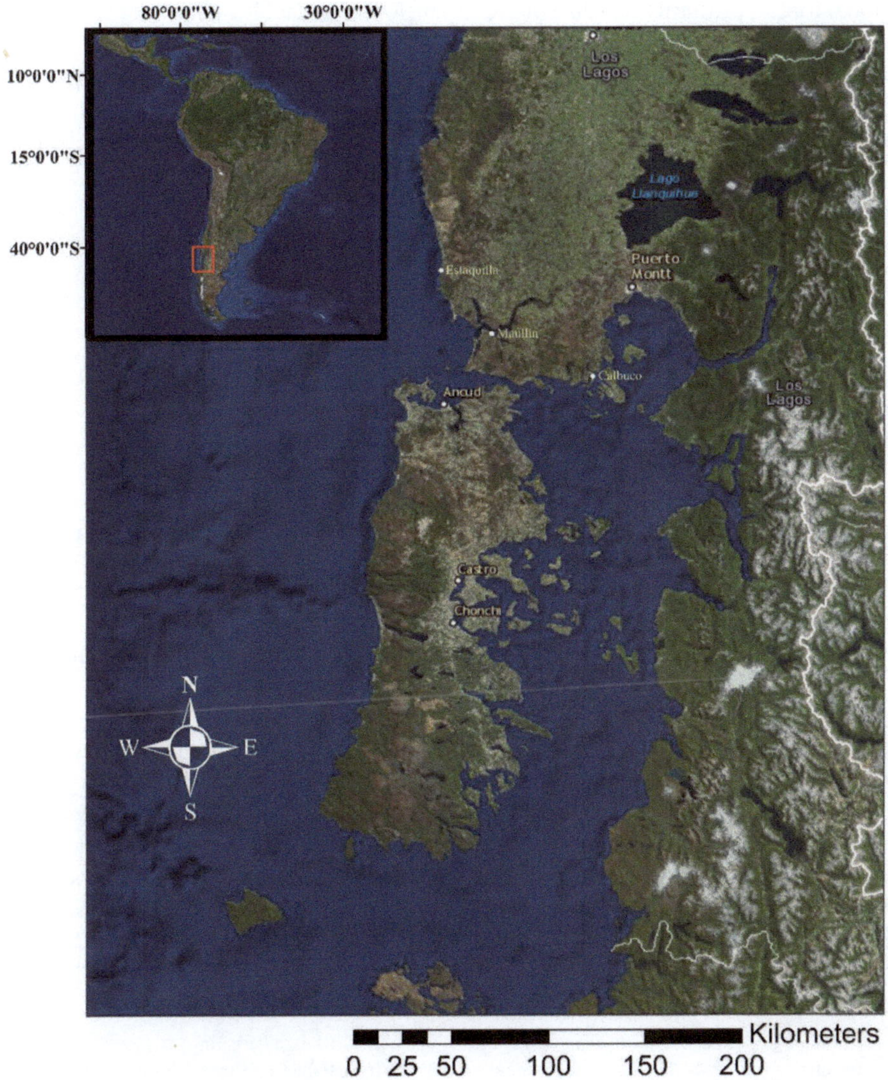

Fig. 3.1 Map of the coastal Lakes Region and the communities of Estaquilla, Carelmapu and Maullín, Calbuco, and Ancud. (Map credit: Sarah Ebel)

Estaquilla, Carelmapu and Maullín, Calbuco, and the community of Ancud, on the big island of Chiloé (see Fig. 3.1). These communities are situated one to two-and-a-half hours from Puerto Montt, the region's capital and the province's only major city. Rolling green hills, ideal for sheep and cow pasture, run down to dramatic, steep bluffs lined by sandy beaches. Intense winds and heavy rainfall, especially in the winter during the *loco* season, make launching and navigating fishing vessels a challenging task. Increasingly, fishers are caught in interactions between local and global forces, primarily because of the expansion of aquaculture farming and global aquaculture exports.

Many fishers today struggle to make a profit due to increases in vessel costs, low prices for product, and vulnerability from stresses in the environment. Increasingly, aquaculture farms are encroaching upon fishing areas. As a consequence, fishing families often find that they have to transition from artisanal fishing to work as divers or processors for aquaculture companies (Pitchon 2015). They recognize the irony of being forced out of fishing to work for the companies that they perceive are killing their livelihood, and many of them are hesitant to leave their professions because fishing allows them to maintain their social connections and feed their families (Pitchon 2015).

Fishing: A Way of Life

In the Lakes Region, fishers are not only economically dependent upon fishing. They also identify with the occupation, and their social and cultural lives are inextricably tied to the craft. When I asked fishers why they chose fishing as an occupation, many said fishing gave them a sense of freedom and a connection to their communities. They also fished out of necessity, and for some, the craft became their passion.

In 2016, I visited Don Simon Dias, a fisher in Calbuco, a community close to the city of Puerto Montt on the mainland. There are around thirty thousand inhabitants in the town and its surrounding areas. Hills gently descend to inlets and coves, and off Calbuco's shore are several islands which add geographical complexity to the area, making it ideal for salmon aquaculture. Aquaculture firms have set up many salmon pens in the area, with large enclosed nets for salmon. These pens encroach on fishers' harvesting areas, causing tension and creating a contested space. Despite the impinging farms and availability of work outside of commercial fishing, ties to fishing remain strong, particularly for older fishers.

> When I asked Don Simon why he fished, he said,
>
> That's a good question, very good question. Why do I fish? Fishing, it's part of our psychology, our social lives, our sustenance. We are free, fishers are free. You leave for the sea, and if you want to return, you return; if you want to work, you work. You have liberty in your decisions…there is a social aspect of working as a shellfish diver. We work in distinct places, and in the evening, the whole *caleta* gets together to share food, converse, and joke with each other. The life of a fisher is very beautiful.

Doña Rosa Toro, a fisher from Estaquilla, a town to the northwest of Calbuco, shared a similar sentiment.

> Our whole family is dedicated to the sea, in different forms, but it is what we do: to go fishing with nets, diving, working along the sea. In a way, everything we do is there. I fish because I like to fish, it is also out of necessity—diving is the source of my livelihood, and I fish for my family.

Estaquilla is a rural community with a population of around five hundred people, located 2 h from Puerto Montt. The last few kilometers of the road into the town were paved in 2014, but the pavement still does not reach the *caleta*. The road winds down hills of lush, green trees toward the ocean. As the road approaches the *caleta*, there is a small cemetery on a hill, and a couple of small stores, called *negocios*. A few churches and small houses dot the hillsides, with cattle grazing among them. Estaquilla has no major industry besides artisanal fishing, and it is common to see pick-up trucks parked outside the town hall or a church, indicating a fishing union meeting.

During the fishing season, small wooden open-boats line the beach. Some individuals move from their houses above the *caleta* to small shacks on the beach in the summer for easy access to their boats during the sea urchin, crab, barnacle, and seaweed harvesting season, depending on environmental conditions. The shacks have one room, a small woodstove, and a camp stove for cooking. One fisher said he loves living along the *caleta* during the fishing season; it makes him feel close to his work and he can rely on the sea for his family's sustenance. At the end of the summer fishing season, they move back to their homes in town and fish for *loco* from May to August, and work in subsistence vegetable farming and raise livestock. Fishing, they maintain, is the most important of all to their well-being. He said they depend on their marine harvest, not only for their income but their diets, family ties, and social lives.

Seasons for *loco* were instituted as part of the fisheries and aquaculture law, *La Ley de Pesca y Acuicultura* for small-scale fishing. The policy was formed after overexploitation of the species during a time of open markets and increased globalization in the 1970s and 1980s (see Chap. 1). After extensive studies on the *loco's* natural life history and human-*loco* interactions in central Chile (Castilla and Fernandez 1998; Castilla and Gelcich 2008), the policy gave fishers exclusive, non-transferable access rights to specific territories, which were to be managed by local fishing unions. The law tied fishers to specific areas for diving, while the studies of marine animals' natural life histories determined the season in which they now fish. Outside of the *loco* season, fishers harvest other benthic resources such as sea urchin, clams, crab, and barnacles, in open-access areas outside of their defined territories. Clearly defined fishing seasons helped stabilize the market for the selling and export of fishery products, which increased profit for many fishers (Rosas et al. 2014). Furthermore, it allowed fishers to maintain diverse seasonal livelihood strategies while simultaneously developing their specialized and intimate knowledge of the inshore marine environment close to their *caletas* (Image 3.1).

Just south of Estaquilla, across the Maullín River, are the communities of Maullín and Carelmapu. The town of Maullín and its surrounding areas are home to around fifteen thousand people. There is one small supermarket, a small bus station, a hospital, and houses clustered into several sectors. Maullín has a small tourist economy, in part because of its proximity to Puerto Montt, its accessibility by a main road, its transportation infrastructure, and its location on the Maullín River, a tourist destination for trout fishing. Despite Maullín's slightly more diverse economy, it remains a port town and reliant on fishing. Don Roberto Molina, a fisher from Maullín explained,

Image 3.1 Fishing vessels in the *caleta* of Estaquilla. (Photo credit: Sarah Ebel)

> It [fishing] is a job for the people, it is an economic well-being for people, and it is very strong here. It is very profitable for people, but of course the risk is very great. The risk is very big, but you get used to it, you learn to know nature. Fishing, therefore, is done as a kind of family bond. It is the way of the sea, that you know when it is bad on the sea, you recognize when it is not good to go to work, or when you can work by the tide. Then you gain confidence, and as you continue, then, it's like any job. Then you realize it's normal work, like anything, like when you're working in the office…you become accustomed to it. They taught me this [fishing] since I learned to walk, and continued to teach me until I was 23 years old. At first, I was afraid of the sea, to go out so many kilometers in the ocean. Then, I got used to it, except for the dizziness, I have not yet overcome that. But all of that is the work of a fisher, who works all day on the sea and does so out of necessity, for his labor, for his chosen lifestyle.

The community of Carelmapu is a short drive from Maullín. Home to only three thousand people, it is accessed by a narrow road that ends at the tip of a peninsula. The roads are unpaved, and there are a few small *negocios* and a small shellfish processing plant. There are few opportunities for work in Carelmapu outside of fishing, and many people who do not want to or cannot work on the sea move away. Don Cristian Auenante, a fisher from Carelmapu acknowledged that he started fishing because he had no other option, but that what began as a necessity morphed into a passion.

> When I began diving, I had no other option. After I started, I was captivated by the sea. It's like, I have two grand passions in my life outside of being a father: to live and preach a Christian life, and to dive for shellfish.

However, Don Jamie Gonzales, a fisher from Ancud, a larger community on the rural island of Chiloé, stated that he believed most fishers fish out of necessity,

> Most of the fishers, and in my case- we are fishers out of necessity, not out of conviction, not because we want to be fishers, but because we were born near the sea, we grew up there, and we have no other option than to fish. We do not have the means to continue studying, others did not have the means from their parents for a good education, then there was nothing else to do than to go to sea.

Ancud is home to more than thirty thousand people, and its economy was once based in artisanal fishing. Now, the economy is more diversified because of aquaculture farms and the large number of tourists who visit during the summer months. A ferry and several bus companies service the big island of Chiloé, picking up individuals in the bus station in Puerto Montt and dropping them off at the bus station in Ancud. Similar to Calbuco, large aquaculture farms take up significant sections of the ocean area surrounding Chiloé. Declines in marine harvests and competition for ocean space from the farms have forced many fishers in the Ancud area to leave the craft of artisanal fishing and pursue work with the aquaculture companies as divers or processors (Pitchon 2011). Nevertheless, artisanal fishers persist, still contributing significantly to the island's economy and workforce.

The craft of artisanal fishing in the Lakes Region holds economic, cultural, and social value for fishers, their families, and their communities. However, impending and possibly irrevocable changes in the ocean's ecosystem and climate, as well as the proliferation of aquaculture, loom over fishers' livelihood security.

A Way of Life Threatened

There are many challenges to making a living as a fisher. Globalized markets, declining fish populations, poor governance of marine resources, and habitat and climatic changes threaten fishers' ways of life and the source of their livelihoods (Beddington et al. 2007; Béné 2003; Béné and Friend 2011). For fishers in Chiloé and the Lakes Region, these threats are becoming increasingly severe, altering fishers' relationships with the marine environment. Despite fishers' love for the craft and fishing's role in their economic and social well-being, there was a palpable sadness in how they spoke about the present and the future of fishing. Fishers said that the biggest threat to their livelihoods is the salmon-farm industry, explaining that they must fight for ocean space. They also shared their perceptions of ecological change, which fishers believed were compounded by poor practices in salmon farming and the complicity of the government. They were suspicious of the government's and biologists' declaration of red tide, and said they did not trust the government or the aquaculture companies. They suggested that the crisis of 2016 (see Chap. 1) was just the beginning of things to come, and they were urging their children and grandchildren to pursue other types of work.

Don Simon Dias from Calbuco said,

> I think the salmon industry in this moment is the problem. They knew it [red tide] was going to happen, but not so quickly. Imagine the mortality of the salmon. They should not get to produce so much fish if they do not have an emergency mechanism in place in case the salmon die. Then, imagine the reaction of the sea. The state and the salmon industry are complicit. They say it is a red tide, but it never was a red tide. They wanted people not to eat contaminated seafood, they shut down our fisheries. But then, some people tried eating the seafood, and nothing happened. They were fine.

In this statement, he explained that a red tide would contaminate seafood so it would be poisonous to human consumption. However, a few individuals in Calbuco ate shellfish and showed no immediate or obvious health consequences. It made the fisher skeptical of the government's and biologists' declarations of a red tide, and he said he lost trust in the government. I heard of similar experiences from fishers in Estaquilla, Carelmapu, and Ancud. This, among other anecdotal evidence from fishers, including stories of individuals who saw the aquaculture companies dump infected fish near the shore of Chiloé, led fishers to believe that the alleged red tide was a farce. Instead, they said it was an infection from the salmon farms that caused the marine die-offs. This remains unproven, but fishers' perceptions that the government is untrustworthy may have repercussions for any future relationships between the government and the fishers.

This ecological devastation, and its subsequent social impact, brought into question future opportunities for artisanal fishers in the Lakes Region. Many fishers I spoke with said they came from several generations of fishing families, and when I asked them if their children fished, most fishers said no. Don Simon said sadly,

> We grew up with the sea, but there is no future in fishing. My children still study topics related to the ocean in university, but they do not fish.

Don Jamie Gonzales from Ancud explained his hopes for his children,

> My son, he just finished school last year. Then I tell him, study, try to study something, I'm economically a little better, I can give him the opportunity to study right now, and I tell him that he could continue studying, but I can't tell him what to do. He does not understand yet - 17 years old - I said, I hope he studies something, for example: something about the sea, marine biology, try to protect our product, to keep looking after this beautiful island, but it will depend what he wants to do.

He added,

> We did not want the salmon farms to enter our area. We had a great fight with the salmon farms and told them that we do not want them to come and pollute our waters. Now the "red tide" has arrived, that was what hurt me the most. I thought that someday, at least, what I said would be recognized because I have been a strong advocate of taking care of our waters. The people come here, tourists or someone who wants to come to eat the product of Ancud and the products of Chiloé, the products are now contaminated. If you go to Castro, Chonchi, Queilen, there, forget it, all products are contaminated, because the salmon farms throw away excess antibiotics and all that is below the pens are contaminated, dead…I believe that in fifteen years, artisanal fishing will not exist to such an extent. We are already old, I have about 50 years, and in ten years more I will have 60. In ten years more, I will no longer work on the sea. And if there are no more people who really care for the sea… today people don't care - our water touches the big companies, and now - I do not know how we are going to live. People are going to be all workers of the companies.

Don Roberto Molina from Maullín reiterated others' concerns, and suggested that the government was not interested in caring for the marine environment,

> I would really like people from outside to do research, to visit this country and save these unique environments that remain in this world - look at it - they are killing it, aquaculture is killing it - look here, this was a sanctuary, a place of nature. It was a whale sanctuary. Nowadays, krill is being threatened by all these salmon companies, that food, they are taking possession of all this inner water that exists in the Southern Cone, they are killing all these mammals that exist in the bottom of the sea. It is something incredible daughter, but it is the pure holy truth. Chile is killing its sanctuary. And that is the reality. There is no other explanation more about this subject.

The reality of environmental change and the influx and proliferation of aquaculture farms is a poignant one for fishers in the Lakes Region. Fishers' abilities to cope with, or adapt to, uncertainty and the ramifications of abrupt changes are of upmost importance in maintaining their ways of life.

Livelihood Diversification as Resilience to Abrupt Environmental Change

It is well known that artisanal fishers around the world engage in a range of livelihood strategies to cope with uncertainty or abrupt economic, ecological, or political change (Allison and Ellis 2001; Goulden et al. 2013). Fishers in the Lakes Region diversify their fishing livelihoods by working in other businesses, including subsistence vegetable and livestock agriculture, and the peddling of shellfish, firewood, and excess vegetable produce. Such diversification has allowed some individuals to stay in their communities despite the hardships of fishing. Others, including many young people, are unable to find or create work, and have either become divers for aquaculture companies or have left their rural communities to seek wage labor elsewhere.

Don Roberto Molina from Maullín diversified his livelihood by working with his father-in-law to operate a boat company. The boat brings tourists and individuals who need to commute between Maullín and the community of La Pasada across the Maullín River. He said that this type of work earns his family extra money, and makes him and his family more resilient to uncertainty in fishing:

> The other alternative that I have is that I work for a ship company. I have that privilege, you see, I take people from one bank to the other. It helps us to generate new income for us in bad weather because the sea is not always good. Others are not so fortunate as we.

Across the strait from Maullín on the big island of Chiloé, Don Jamie Gonzales spoke with me about the environmental devastation of 2016 and his worries about being the sole income earner for his family of four. With a teenage son and a new born baby, he said his need to support his family weighed upon him. He expressed gratitude that he had the opportunity to study, but that although he studied at a trade school in Santiago, he returned home to go fishing. In our conversation, his gratitude for the opportunity to study conveyed a sincere relief that if fishing were no longer an option, he would have another way to provide for his family, even if it was not a lot of money.

> I studied - thank God, I studied. I had the option to study but I did not like it - that is, I did not have the opportunity to work in it either, because the salaries are so low. I am a mechanical technician, I studied in Santiago, and later I could not study engineering because I did not have the money to pay for school. From there [engineering] I would have had a higher salary, but the mechanical technician is very poorly paid, so I returned to fishing.

He then looked at his newborn son and said, "He will never know fishing."

South of Puerto Montt, the community of Calbuco is dependent upon fishing and subsistence agriculture. Don Simon and his wife showed me their extensive vegetable gardens and their greenhouse, built to produce more food during the winter months. Although the region near the coast does not typically have frosts in the winter, the heavy rains and wind can harm growing plants. Initially, the gardens and the greenhouse were a way for them to grow food for their family. More and more, they sell produce for supplemental income. However, the household's ties to agriculture were not as strong as their ties to fishing. When I asked them about their greenhouse and farming, they said it contributed to security for their family and neighbors and could help make a small amount of income if they needed. Fishing, they said, provided nearly all of their income. Moreover, their social life in Calbuco was inextricably linked to fishing. He said he and his wife would gather with other fishers and their families to share food and company many nights during the week. Outside of fishing, they were involved in the church.

Don Simon and his wife raised five children in their small home, and their children now work as engineers, teachers, or are studying in university. While their children were growing up, their family depended on fishing, but he said times have changed. He said it is "no longer profitable for his children to dedicate their lives to the profession of the fisher," and that it is getting harder and harder to make ends meet while they are reliant on subsistence agriculture while the fisheries are closed (Image 3.2).

In Estaquilla, Doña Rosa Toro lives in a small, bright red and blue house, 2 h from Puerto Montt. Green pastures surround her house and sheep graze among the grasses. She and her husband, both fishers, raised four children, all of whom now make their living on the ocean. Her husband passed away in 2012, and her son took his place in his fishing union and dives for *loco* during the open season. He also works as a merchant marine for a large industrial marine logistics and shipping company. The three other children, all young women, work as merchant marines and as commercial divers for aquaculture companies. They are independent and hardworking, and are not new to hardship or to coping with changes in the ecosystem, economy, or politics. When fisheries were closed in the late 1980s because of overharvesting, she and her family experienced extreme poverty. Rooted in their relationships with the environment, they coped by cutting and selling firewood and raising livestock. Two of the daughters hope to attend university in Puerto Montt,

Image 3.2 Sheep grazing near a fisher's home in Estaquilla. (Photo credit: Sarah Ebel)

but the expense of higher education proves to be a significant barrier to furthering their education.

To supplement their income during the red tide crisis, Doña Rosa would drive the 2 h to the city of Puerto Montt with cut and bagged firewood many times a week. I went on several of these deliveries with her and one of her daughters. When we returned home from one trip, the mother hopped out of the truck and disappeared over a hillside. Within moments, she was driving her four cattle back to the small barn for the night. She planned to sell the cattle and lambs at Christmas time. Her knowledge of small-scale farming and livestock-raising was impressive; she was the only fisher I spoke with who raised livestock other than chickens, and who had such an intimate relationship with farming. She said she enjoyed farming, but that it was hard work. Moreover, cutting wood and raising lambs to sell during the Christmas season helped her financially, but was not enough to live on:

> With the firewood, you cannot maintain your life. But if you go fishing, you can sell the fish and buy things you need for your house, or your vehicle. The wood is not enough to pay the expenses. Nothing but fishing brings in enough for that. Selling the lambs also brings in too little, but they help. Apart from being fishers, we have to be small farmers.

Although Doña Rosa was a farmer out of necessity, she said her ties to the sea remained strongest. Her day-to-day finances depended on their harvests from the sea, and her preferences for her and her family's diets were also rooted in the sea. On most nights, one found *mariscos* (shellfish) or finfish with potatoes or bread on the dinner table in their household.

During the crisis of 2016, the family could not dive for subsistence. My conversations with them at dinner during the winter of 2016 often revolved around the lack of *mariscos* in their meals, and they relied on neighbors who harvested finfish. Compounding the problem of no subsistence fishing was the difficultly they encountered in paying for food and other household costs. One of the daughters moved home for the winter to help her mother with the farming, while the other children contributed financially from their jobs away from Estaquilla.

Fishers in the Lakes Region used their diversified livelihood strategies to cope with the fishery closures during the abrupt environmental crisis of 2016. However, coping—conceptualized as reactive adaptations used for short-term survival (Fabricius et al. 2007; Smit and Wandel 2006)—may not be enough to foster individuals' resilience to ecological, economic, or political stresses in the long term (Fabricius et al. 2007; Marschke and Berkes 2006), leaving fishers in the Lakes Region vulnerable to poverty.

Limitations of Livelihood Diversification

"I am one of those who claimed, many years ago, that the salmon companies sooner or later would kill our resources. What's more, this [environmental event] is a warning, a beginning of a mortality that has come in time, because this is not going to stop, daughter, it will not go away. It will not stop, unless the government awakens and knows how to understand that sooner or later, these companies are going to kill our ecosystem, they are going to kill us."—Don Roberto Molina from Maullín, 2016.

Strategies to supplement income are practiced by many fishers around the world. These strategies are sometimes considered coping mechanisms, when the individual reacts to some form of stress that has impacted the social, economic, or ecological system (Fabricius et al. 2007). In Chiloé and the Lakes Region, fishers relied on their seasonal livelihood strategies to cope during the abrupt crisis of 2016. Yet, livelihood diversification was not enough for fishers to maintain their ways of life. Fishing provides the majority of people's income in this part of the region. With the closure of wild fisheries during the crisis, many fishers struggled to feed their families despite engaging closely with small-scale agriculture. Moreover, they felt that they lost their social and cultural connections to their neighbors and to their history. Without fishing, individuals organized fewer social gatherings outside of church activities. Fishers' children also had to move away to find work, ending a generations-long tradition of fishing for many families in the region. These repercussions from the 2016 crisis illuminate the limitations of livelihood diversification as a form of resilience for individuals whose well-being is embedded in the natural environment.

Reacting to stresses or relying on seasonal work, as did the fishers in the Lakes Region, may only ameliorate an individual's situation for a brief time, and can leave the individual vulnerable to future stresses or long-term change (Fabricius et al. 2007). Artisanal fishers who live in rural and impoverished areas, like the fishers in the Lakes Region, often lack the resources that are needed to foster adaptation or sustain long-term resilience. These resources, which include formal education, financial capital, healthcare, and transportation, are vital in maintaining the resilience of social, economic, and ecological systems (Fabricius et al. 2007; Goulden et al. 2013). It is only with resources such as these that individuals can adapt effectively to maintain their livelihoods over the long term (Fabricius et al. 2007).

Future analyses of individual's resilience must move beyond questions of livelihood diversification to address larger questions of ocean governance, accountability, and adaptive management of natural resources. In 2016, fishers called upon the government for accountability and for change to policy, and they have yet to receive a clear answer. Don Roberto Molina from Maullín said sadly,

> This is unfortunate for us as artisanal fishermen, I see Chile as one of the—you could say, as a power in the subject of what is fishing, because there is still natural fishing, artisanal fishing. We want to try to save it in time, but we need the support of the government, the government that will help us save these wild fisheries, so that it can survive for many more years if we want, if the government or this country would like to keep this fishery for many years. The problem is that there is no interest in doing so.

References

Allison, E.H., and F. Ellis. 2001. The livelihoods approach and management of small-scale fisheries. *Marine Policy* 25 (5): 377–388.

Barrett, G., M.I. Caniggia, and L. Read. 2002. "There are more vets than doctors in Chiloé": Social and community impact of the globalization of aquaculture in Chile. *World Development* 30 (11): 1951–1965.

Barton, J.R., and A. Fløysand. 2010. The political ecology of Chilean salmon aquaculture, 1982 2010: A trajectory from economic development to global sustainability. *Global Environmental Change* 20 (4): 739–752.

Beddington, J.R., D.J. Agnew, and C.W. Clark. 2007. Current problems in the management of marine fisheries. *Science* 316 (5832): 1713–1716.

Béné, C. 2003. When fishery rhymes with poverty: A first step beyond the old paradigm on poverty in small-scale fisheries. *World Development* 31 (6): 949–975.

Béné, C., and R.M. Friend. 2011. Poverty in small-scale fisheries old issue, new analysis. *Progress in Development Studies* 11 (2): 119–144.

Castilla, J.C., and S. Gelcich. 2008. Management of the loco (Concholepas concholepas) as a driver for self-governance of small-scale benthic fisheries in Chile. *FAO Fisheries Technical Paper* 504: 441.

Castilla, J.C., and M. Fernandez. 1998. Small-scale benthic fisheries in Chile: on co-management and sustainable use of benthic invertebrates. *Ecological applications* 8 (sp1): S124–S132.

Ellis, F. 2000. *Rural livelihoods and diversity in developing countries*. Oxford: Oxford university press.

Fabricius, C., C. Folke, G. Cundill, and L. Schultz. 2007. Powerless spectators, coping actors, and adaptive co-managers: A synthesis of the role of communities in ecosystem management. *Ecology and Society* 12 (1).

Goulden, M.C., W.N. Adger, E.H. Allison, and D. Conway. 2013. Limits to resilience from livelihood diversification and social capital in lake social–ecological systems. *Annals of the Association of American Geographers* 103 (4): 906–924.

Holling, C.S., and L.H. Gunderson. 2002. Resilience and adaptive cycles. In *Panarchy: Understanding transformations in human and natural systems*, 25–62.

INE. 2008. Instituto Nacional de Estadisticas de Chile. www.ine.cl.

Latta, A., and B.E.C. Aguayo. 2012. Testing the limits: neoliberal ecologies from Pinochet to Bachelet. *Latin American Perspectives* 39 (4): 163–180.

Marschke, M., and F. Berkes. 2006. Exploring strategies that build livelihood resilience: A case from Cambodia. *Ecology and Society* 11 (1).

Moreno and Revenga. 2014. *The Systems of Territorial User Rights in rights in Chile*. Arlington, VA: The Nature Conservancy.

Peterson, N.D. 2014. Breaking the bounds of rationality: Values, relationships, and decision making in Mexican fishing communities. *Conservation and Society* 12 (3): 245.

Pitchon, A. 2011. Sea hunters or sea farmers? Transitions in Chilean fisheries. *Human Organization* 70 (2): 200–209.

———. 2015. Large-scale aquaculture and coastal resource-dependent communities: Tradition in transition on Chiloe Island, Chile. *The Journal of Latin American and Caribbean Anthropology* 20 (2): 343–358.

Rosas, J., J. Dresdner, C. Chávez, and M. Quiroga. 2014. Effect of social networks on the economic performance of TURFs: The case of the artisanal fishermen organizations in Southern Chile. *Ocean & coastal management* 88: 43–52.

Smit, B., and J. Wandel. 2006. Adaptation, adaptive capacity and vulnerability. *Global Environmental Change* 16 (3): 282–292.

Urquhart, J., and T. Acott. 2013. Constructing 'the Stade': Fishers' and non-fishers' identity and place attachment in Hastings, south-East England. *Marine Policy* 37: 45–54.

Walker, B., C.S. Holling, S. Carpenter, and A. Kinzig. 2004. Resilience, adaptability and transformability in social–ecological systems. *Ecology and Society* 9 (2): 5.

Chapter 4
Food and Culture in Chiloé: Potatoes, Curanto, and Chicha

Anton Daughters

Restaurant Travesía sits on a narrow city street bordering Castro's shoreline. Run by Lorna Muñoz, a thirty-something chef and native of Castro, and Renato Cárdenas, a prolific writer and cultural historian, the restaurant specializes in food that is endemic to the region. Fish, shellfish, and dozens of varieties of potatoes form the core ingredients of many of its dishes. "Our meals, like all things on these islands," write Cárdenas and Muñoz in their 2015 book *Chiloé Contado Desde la Cocina*,[1] "are influenced by the two founding cultures of Chiloé: the Mapuche and the Spanish." Renato Cárdenas expanded on this point in an interview in 2017: "Potatoes, fish, shellfish, and curanto, among other things, came from Mapuche, or more precisely, Huilliche traditions. The Spaniards brought livestock—pigs, cattle, and sheep. And they brought new crops, like apples, wheat, and barley. The food and traditions of this archipelago are a fusion of those cultures."[2]

In the following pages I explore the cultural significance of potatoes, *curanto*, and *chicha*, food that Chilotes uphold as significant markers of islander identity.[3] The traditions and beliefs surrounding these products are reflective of the unique elements of Chiloé's past and present. They paint a picture of a rural way of life that most islanders hold in high esteem, and they are emblematic of the broader-based

---

[1] *Chiloé As Told From the Kitchen.*

[2] Interview with Renato Cárdenas, Castro, Jan. 2017.

[3] The information in this chapter is drawn from participant observation fieldwork and interviews carried out by the author between 2006 and 2017, supplemented by primary historical sources and scholarly articles.

A. Daughters (✉)
Department of Sociology, Anthropology, and Justice Systems, Truman State University, Kirksville, MO, USA
e-mail: adaughters@truman.edu

A. Daughters, A. Pitchon (eds.), *Chiloé*, Ethnobiology,
https://doi.org/10.1007/978-3-319-91983-6_4

values that Chilotes emphasize. As Muñoz and Cárdenas assert, islander culture is perhaps best understood through its food.

Potatoes

Nearly every European explorer, military official, naturalist, or castaway to pass through Chiloé from the late 1500s through the early 1900s commented on the abundance and delectability of potatoes. Writing around 1670, the Jesuit Diego de Rosales observed that "in Chiloé, the entire subsistence of the natives comes down to some roots of the earth that they call potatoes (*papas*)… of which they harvest large quantities in order to have enough, and serve in place of bread" (Rosales 1878, translation mine). The British castaway John Byron, recovering in Castro from his arduous journey along the Patagonian coast in 1742, wrote that Chilote households enjoy "plenty of the finest potatoes in the world: these are always roasted in the ashes, scraped up, and served at meals instead of bread" (Byron 1812: 142). Visiting Chiloé's big island in 1834 as part of his voyage aboard the H.M.S. Beagle, Charles Darwin commented on the astonishing array of wild and domesticated potatoes found throughout the archipelago (Darwin 1998). He shipped several varieties back to England to help him understand the biological concept of "species" and to investigate how plant pathogens cause blights (Ristaino and Pfister 2016).

The potato (*Solanum tuberosum*) is a remarkable species in many respects. All varieties originate from one of two regions in South America: the Andean highlands of Bolivia and Peru or the Patagonian coast of Chile, including the islands of Chiloé. Chiloé alone boasts hundreds of varieties farmed by indigenous communities prior to the arrival of Europeans. At the archaeological site of Monte Verde, located on the South American mainland just north of Chiloé's big island, researchers have uncovered traces of wild potatoes dating to 14,800 BP (Dillehay et al. 2008). Evidence of human consumption of potatoes dates to more than 12,000 BP (Dillehay 1989). Although researchers continue to debate the ultimate origins of the species, recent genetic studies indicate that more than 99% of the potatoes cultivated worldwide today trace their origins to domesticated varieties from Chiloé that were exported to Europe sometime between the seventeenth and nineteenth centuries (Ames and Spooner 2008; Yao 2010).

Islanders often boast about Chiloé's role in the meteoric rise of this important food source. Easy to grow and packed with nutrients, potatoes have become the world's fourth-largest crop, behind corn, wheat, and rice (Nabhan 2016). "Chiloé has helped feed the world," I was told by Renato Cárdenas in 2006. The farmers on the island of Llingua, with whom I spent considerable time between 2006 and 2017, would comment on the global appeal of their crop, often in understated tones, but with a note of pride.

The Mapuche word for potato is *poñi* (Cárdenas 1996). A staple of Mapuche communities on the mainland and Huilliche settlements on the archipelago, potatoes were the focal point of the agricultural work carried out by indigenous groups prior to the arrival of Europeans. Despite the introduction to southern Chile of non-native crops from the Old World in the sixteenth and seventeenth centuries, including wheat—which was widely adopted by islanders—potatoes maintained their preeminence in Chiloé as the most commonly farmed crop, with varieties ranging in color from deep purple to yellow and red. Over the years, idiosyncratic folk beliefs, unique recipes, and everyday work patterns developed around the cultivation of this staple food (Cárdenas and Villagrán 2005).

Common Beliefs About Potatoes

The most widely documented folk beliefs, described by Renato Cárdenas and Carolina Villagrán in their 2005 overview of the botany of Chiloé, reflect a deep reverence for the crop. Rural islanders will often say that one must never till the earth with anger when planting potatoes, otherwise the harvest will get ruined. Carrying a basket full of potatoes across a cemetery will leave the potatoes "smelling of death" (Cárdenas and Muñoz 2015: 39). Forgetting that you have put potatoes in the smokehouse or the oven, such that they burn, might lead to your family forgetting who you are when you are older. And when planting the first potato of the year, as well as eating the first potato crop of the year, you must always make the sign of the cross.

Potatoes are also seen to have medicinal value by many islanders. While carrying out fieldwork on the island of Llingua in 2006, I was fed soup consisting primarily of boiled potatoes to combat a cold I had been struck with; my host insisted that it would help me recover, and indeed it did, or at least appeared to. I was also told that potatoes helped with arthritis and inflammation, a claim that seems to be backed by recent scientific research (Kaspar et al. 2010). Islanders place slices of raw potatoes on their skin to help with headaches, fever, and burns. On some islands, internal hemorrhages are treated with water from boiled potatoes; the ailing person is to drink the water and also bathe in it. And carrying a potato in your pocket at all times will, according to an old Chilote belief, help prevent general inflammations, injuries, and even ward off spells from spiteful neighbors (Cárdenas and Villagrán 2005).

A variety of recipes that are unique to the region reflect the centrality of potatoes in the everyday diet of Chilotes. The very words for these foods have become synonymous with Chiloé itself: *milcao*, *chapalele*, *tropón*. Many of these recipes are simply potato patties prepared in different ways. *Milcao*, for example, are grated potato patties stuffed with salted pork that are either baked or fried in lard; *chapalele* are boiled, mixed with flour, and generally baked in the ground under a layer of *nalca* leaves; and *tropón* are freeze-dried and typically grilled over hot embers (Jiménez 2009). All three patties can be prepared as part of a *curanto*, a traditional dish of shellfish, potato patties, and chicken that I will discuss later in this chapter.

La Minga

Beyond these everyday recipes and folk beliefs, the potato has played an essential role in the broader development of Chilote culture, namely patterns of labor and social relations. The reciprocal labor tradition known as the *minga*, also discussed in Chaps. 1 and 5 of this volume, developed largely as a consequence of the need for labor in Mapuche and Huilliche agricultural communities. Potatoes were the central crop for the Huilliche; therefore, families often needed help from neighbors to plant and harvest their potato fields. Labor received from a neighbor meant labor owed to that neighbor. Over time, families developed strong networks of reciprocity throughout their communities—sets of neighbors and extended family with whom they were mutually indebted (Cárdenas Álvarez et al. 1991).

These patterns of labor remained in place even after the arrival of Europeans to Chiloé in the mid-1500s. Chiloé was a poor, relatively isolated colonial province, sandwiched between sparsely populated Patagonia to the east and south, and Mapuche-held lands to the north. Very little cash circulated throughout the archipelago, and supplies from the larger Spanish settlements in the north arrived sporadically, sometimes once every several years. As a consequence, indigenous practices like the minga were widely adopted, remaining commonplace on the archipelago through the independence wars of the nineteenth century and even the industrialization of the big island in the late twentieth century (Daughters 2016b). Assisting one's neighbor with labor became a deeply ingrained cultural imperative. An ideal neighbor expressed his or her *solidaridad* (solidarity) by lending their sweat and muscle.

Mingas are far less common today given that large swaths of Chiloé have become urbanized. However, they are still practiced regularly by agricultural families on rural sectors of the archipelago, especially islands like Meulín and Apiao where potato farming is widespread (see Chaps. 5 and 7). Nearly all Chilotes identify, to varying degrees, with the notion of neighborly solidarity born out of labor reciprocity. Those sentiments are strongest in sectors that still rely at least partly on subsistence fishing and agriculture, but they also exist among the urban youth of Chiloé, many of them raised by parents who were active in mingas. "We are born *solidario*," I was told by a high school student in Achao in 2006 during a wave of national student protests that year.[4] He and his peers went on to explain that many young Chilotes identify with social movements because of the strong history of the minga and reciprocity.

Similar sentiments were articulated during the archipelago-wide protests in 2016 over the red tide and salmon-dumping crisis. Graffiti in urban Ancud and Castro often made reference to islander solidarity as a cultural value that stood in contrast to the profit-driven and environmentally harmful salmon industry. Speeches made by high school students, scuba divers, and fishermen at rallies across the big island in May of 2016 often framed the crisis as a consequence of capitalist interests

[4] Benjamín, Achao, June 7, 2006.

Image 4.1 A potato-planting minga on the property of Hugo and Irene Mansilla in 2006. (Photo credit: Anton Daughters)

encroaching on a culture of mingas, reciprocity, and small-scale subsistence livelihoods (Pfeiffer 2016). In the eyes of many protesters, the salmon industry posed more than an environmental threat: it posed a threat to core elements of islander culture. The biggest symbol of that culture was the minga (Daughters 2016a, b).

Over the years, I participated in mingas on Llingua, Chulín, and Chiloé's big island. On the island of Llingua, where I formed my strongest contacts, I helped Hugo and Irene Mansilla with the planting and harvesting of their potato fields. The work was typically done with assistance from several neighbors who had been helping them for decades, including Irene's brother Aristide, who generally led the oxen. As the fields were plowed, the rest of us placed potatoes in the earth and sprinkled fertilizer over them. We were always well fed, and during our work breaks we were offered hard apple cider (*chicha*). These mingas constituted a form of *días cambiados* (days exchanged): a day of labor received meant a day of labor owed (see Chap. 5 for more on *días cambiados*) (Image 4.1).

More ambitious work projects, like the construction or moving of a house, required larger gatherings. These general mingas, much less common today than *días cambiados*, could involve dozens of neighbors, all working toward a common goal over the course of several days. All participants would expect to be well fed by the hosts, but they understood that their labor would not always be reciprocated in a tit-for-tat manner; rather, a general sense of obligation was imparted onto the hosts. In this regard, these mingas represented a more generalized form of labor reciprocity.

Altogether, potatoes shaped numerous aspects of islander life. They gave rise to folk traditions, had medicinal value, and formed the basis of some of the most autochthonous Chilote meals. The need to plant and harvest potatoes drove the practice of the minga, which in turn strengthened social networks in rural communities. As the most important crop in the region for millennia, its growing cycle dictated work patterns, giving broader shape to rural livelihoods. It is no surprise, therefore, that Chilotes uphold the potato not just as an ordinary crop but also as a food that in many respects symbolizes their way of life.

El Curanto

In 1984 I spent a week in Chiloé with my parents. I was 13 years old, and it was my first time visiting the archipelago. We stopped for several nights in Quellón—a town of about 3500 at the time.[5] On the day we were scheduled to start driving north again, we took part in a public *curanto* feast held near the town's central dock. Several dozen people had gathered, many of them to help arrange layers of *nalca* leaves, fish, shellfish, potatoes, and chicken in an earthen pit lined with steaming rocks. I remember the musky ocean breeze, the stinging smoke from the bonfire, and the dull clatter of shellfish cascading into the pit. Slabs of wood were dragged over the pit once the layers of food were in place, and burlap sacks were then draped over the wood to cover any openings. Several hours later I was handed a bowl of smoky and succulent shellfish and potato patties, one of my best childhood meals.

Renato Cárdenas and Lorna Muñoz (2015: 24) write that the curanto "is among the most ancient recipes of humanity." The simple act of digging a pit, lining it with red-hot stones, and cooking shellfish or other foods in it goes back tens of thousands of years. Evidence of the practice is found throughout the world. According to Alston V. Thoms, a Texas A&M archaeologist, "hot-rock cookery" in southern Europe dates to the late Aurignacion (32000–33,000 BP). Those pits "tended to be basin shaped, about 1.5 meters in diameter, and filled with heat-fractured river cobbles" (Thoms 2009).[6] Anthropologist James S. Stuart (2010) writes: "Similar features are found on all continents, but their purpose is seldom clear. The best known *recent* curantos (geologically speaking, i.e. within the last 10,000 years or so) are found in the Americas and in Polynesia." The New England clambake of the northeast United States—a meal of mussels, clams, crabs, and lobsters layered with seaweed and cooked in a pit in the earth—bears a strong resemblance to the curanto of Chiloé. Likewise, the hāngi of New Zealand—a traditional earth-pit Maori meal—and the kālua of the Hawaiian Islands rely on methods and ingredients similar to the curanto.

[5] The 1982 national census lists the town's population at 3500. The population today is more than 25,000.

[6] quoted in Stuart (2010).

Curantos in Chiloé: A Brief History

Archaeological excavations on Chiloé have uncovered evidence of curantos dating back to the first occupants of South America. One site on the northeast coast of the big island, excavated by archaeologists Pilar Rivas and Carlos Ocampo, shows stones in a pit with shellfish remains dating to 11,500 BP. A similar site, on the big island's Gulf of Quetelmahue, shows a carefully arranged curanto pit dating to between 5000 and 6100 BP, while a third dates to 1830 BP (Rivas and Ocampo 2005: 70). These sites are associated with the nomadic Chonos—coastal foragers who occupied Chiloé for millennia before the arrival of the Huilliche (see Chap. 1). Spanish explorers observed Chonos cooking shellfish in this manner in the 1600s and 1700s. Given such evidence in both the written and archaeological record, most scholars believe curantos originated with Chonos and were adopted later by agricultural Huilliches who migrated into the region sometime during the first millennium AD (Rivas and Ocampo 2005, Cárdenas and Villagrán 2005).

By the 1800s, the simple practice of cooking shellfish in a rock-lined pit had transformed into a festive social gathering, an occasion for strengthening social ties through shared food and shared labor. Traveling in the 1850s, the writer and politician Vicente Pérez Rosales (1886: 383) observed that

> [Fish] and those inexhaustible banks of exquisite shellfish of all types that the low tides expose were, along with potatoes and fava beans, the larder that sustained [the Chilotes]. Even the means of preparing those delicacies was purely Indian, from the time of the conquest. In a hole in the ground full of stones heated by a fire are placed shellfish, fish, meat (if there is any), cheese, and potatoes, and without delay, everything is covered with monstrous *pangui* [nalca] leaves, and finished by covering with sod and earth to keep the steam from escaping. A quarter of an hour later, one saw the entire family, with their obligatory escort of dogs and pigs, surround the smoky horn of plenty in which each one put his hand and ate, sucking his fingers, until satisfied.

The Chilean publisher Recaredo S. Tornero, writing in 1872, noted that curantos are "a type of banquet or feast they celebrate in the fresh air, always at the edge of the beach, and very frequently since there is never a lack of pretext: now a wedding, a baptism, a sick person out of danger, a good harvest, finishing a house, a happy return from the mountains, or simply a desire to have a good time" (Tornero 1872: 386, translated by Stuart 2010).

The word *curanto* comes from Mapudungun, the shared language of the Mapuche and Huilliche. Cárdenas Álvarez (1996: 84) points out that the word is a combination of *cura*, meaning "stone," and *antu*, meaning "sun" or "heat." This would suggest a meaning of "rocks heated by the sun." Others have translated the term *kurantu* to "stony ground" (Stuart 2010).

Earth-Pit Versus Pot Curantos

Today, the traditional earth-pit curanto has been largely replaced by curantos prepared in pots (*curanto en olla*). The same ingredients are used. These include various layers of shellfish (mussels and clams typically), meats (chicken, pork, beef, lamb, or sausages), and potatoes or potato patties (*milcao* and *chapalele*). The food is layered with either *nalca* leaves, called *pangue*—the traditional ingredient—or, more frequently today, cabbage leaves. The shift from earth-pit to pot curantos started in the 1960s, according to Cárdenas and Muñoz (2015). A variation of the *curanto en holla* also developed around this time. Referred to as *polmay*, this dish can be thought of as a "simple" curanto: steamed mussels seasoned with garlic and onion and often served with rice.

Earth-pit curantos (*curantos al hoyo*) are still prepared with some regularity but usually as part of public gatherings at the cultural and religious festivals organized annually throughout the archipelago. The most popular of these festivals are the Fiesta Nazareno on the island of Caguach—a celebration of historic Jesuit missions that is held in August and January—and the *Festival Costumbrista* of Achao held in February. According to Chilote writer Herman Galindo Cárcamo (2002), a single earth-pit curanto can feed up to 30 people. The pit should be no deeper than 50 cm. A fire is built in the pit, and "rocks of average size" are placed on the fire. Once the flames have mostly died, the charred wood is carefully pulled out and the food is layered over the stones. This last step must be carried out quickly to avoid losing too much heat from the stones. To best seal the pit, Galindo suggests, washcloths and sod should be placed over it, with *arrayán* branches separating the sod from the food. The pit should remain sealed at least for an hour, preferably two (Image 4.2).

The stones used in earth-pit curantos have special cultural significance. They are selected with care in advance of the feast. Ideally they are about the size of a fist, sometimes bigger. Preference is given to rocks that have been used in previous curantos, and to volcanic rocks, since these do not split when exposed to the intense heat of the fire. Volcanic rocks, moreover, generally retain heat better.

The rocks used in curanto pits are, according to Huilliche legend, the remains of humans who drowned after the primordial battle, thousands of years ago, between two mythical serpents. When Caicai Vilú—the god-serpent of water—defeated Trentren Vilú—god of land—much of the earth was flooded, giving shape to the Archipelago of Chiloé. The humans who were swept into the waters were, depending on the predilection of Caicai Vilú, transformed either into marine animals or rocks (Cárdenas and Muñoz 2015: 25). Those same rocks today are what retain and radiate heat to cook a meal consumed by so many, to help give nourishment and life to living humans. The curanto, in this regard, is a meal that not only has deep historic roots but also an elemental connection to the mythology and folklore of the region.

Image 4.2 Residents of the island of Caguach prepare a curanto during the Fiesta Nazareno in 2012. (Photo credit: Anton Daughters)

Chicha

Whereas potatoes and curantos are food traditions that have been in place for millennia in southern Chile, hard apple cider (*chicha*) is made from an Old World fruit that was imported to Chile in the sixteenth century. Nevertheless, chicha has become one of the most widely consumed alcoholic drinks in the region, a standard brew served at festivals, work parties, and kitchen tables. It is easy to make and store, and like potatoes and curantos, it is linked to patterns of labor and social relations that characterize Chilote culture.

It is important to point out that the word chicha can refer to a variety of fermented drinks found throughout South America. In the Andean highlands of Ecuador, Peru, and Bolivia, chicha typically describes a *corn*-beer brewed thousands of years before the arrival of Europeans. This kind of chicha, still widely consumed today, is part of Andean lore and tradition and is recognized as an autochthonous indigenous drink, a derivation of one of the most important New World crops.

However, in the semiarid north and central sectors of Chile, especially the fertile central valley where vineyards flourish, chicha generally refers to a *grape*-based brew. Described by some as "a poor man's homemade wine," it is often consumed during the Fiestas Patrias—Chile's September 18 Independence Day celebrations (Daughters 2014). As with apples, grapes are an Old World crop imported to Chile in the mid-1500s, making this version of chicha a relatively recent (i.e., within the last 500 years) imported drink.

South of Chile's Bio Bio River—in a region encompassing Concepción, Valdivia, Osorno, the Archipelago of Chiloé, and Patagonia—chicha refers to brewed *apple* cider. Apple trees grow especially well in the cold, damp climate of southern Chile, making this fruit quite abundant. Darwin took note of this in 1834 when he commented on the speed with which apple trees could be grown: "In three years it is possible to have an orchard of large, fruit-bearing trees…. Not above one in a hundred have any seeds in its core" (Keynes 2000: 237). Indeed, today it is common for islanders to have an apple grove in their backyard. Apples are part of the everyday diet of southern Chileans, and even form the basis of regional folk traditions. In parts of Chiloé, for example, a concealed apple is believed to help a man woo a woman, and dreams of green apples are interpreted as a sign of sorrows to come. Placing an apple near the head of an asthmatic child will cure the boy or girl of their ailment. For Chilotes, according to Cárdenas and Villagrán (2005: 225), "the apple *is* chicha."

Precontact Brews

Prior to the introduction of the apple to southern Chile, indigenous groups were brewing chicha from a variety of fruits native to the region: strawberries, Calafate berries, the fruits of the native molle tree and *murtilla* shrub, and the pine nuts (*piñones*) produced by Araucaria trees (Pardo and Pizarro 2005). Early Spanish explorers commented on these drinks. Traveling the region in the 1600s, the Spanish soldier Francisco Nuñez de Pineda described chicha as "the greatest gift" a host could bestow upon a visitor. He revealed a preference for varieties that were "strong, spicy, and well-aged" (Nuñez de Pineda 1973: 55). Writing later that same century, the Jesuit priest and historian Diego de Rosales stated that one of the most important displays of courtesy and hospitality by indigenous groups in southern Chile was an offering of chicha to a guest. "They say it is better than wine… and the delight of all gatherings" (Rosales 1989: 1378).

Apple trees were first introduced to central Chile in the mid-1500s, and as these spread south, apples gradually replaced native berries and nuts as the preferred source of chicha (Castro San Carlos 2016; Pardo and Pizarro 2005). The new fruit thrived and multiplied along the Pacific coastline. The Dominican priest and chronicler Reginaldo Lizárraga described vast orchards of apple trees growing in central Chile as early as 1561, noting that one colonist had an orchard the size of four city blocks (Lizárraga 1935). An anonymous Jesuit writing in Mendoza in the 1780s described apples as not just the dominant fruit of the region but also a significant regional export and the source of most fermented beverages (Draghi Lucero 1940). Clearly, by the eighteenth century, apples had taken over as the fruit of choice.

The Maja de Manzana

Today, the harvesting and mashing of apples to produce chicha is a significant work event in Chiloé called the *maja de manzana*. Majas typically take place in late summer (March) and are considered a type of minga. Neighbors band together to knock apples off trees, gather them in sacks or buckets, grate them to produce pulp, then squeeze the pulp in wooden or mechanical presses to extract the cider. The help received from neighbors is treated as an act of reciprocity that strengthens the networks of mutual assistance between rural households.

The cider extracted from the maja is usually stored in large wooden or plastic barrels. Within a few days it starts to ferment. Sometimes yeast is added to speed up the process. Most families simply allow the cider to sit in the barrels for months. Older cider, therefore, has a higher alcoholic content than cider that was recently produced. In other words, by late winter, families are drinking chicha that has a stronger punch to it.

In 2006 I took part in a maja on the island of Llingua. The day was sunny and cool as the Mansilla family gathered in their small orchard with about a dozen neighbors and friends. They brought long wooden poles—eight feet or longer—that had been cut for the occasion. We spent the morning knocking hundreds of apples off the gnarled trees scattered across their backyard. The work was light and the mood was festive, and my hosts poked at the apple-bearing branches with swift, deliberate motions, sometimes crouching for a better angle, then grinning at their success at dislodging hard-to-reach fruit. The apples were collected, tossed in nylon sacks, and stored in a wooden shed for mashing later that week.[7] Over the course of that year, I was able to partake in the chicha generated from that late summer maja. More importantly, I had lent my labor to the process, strengthening ties with the family and demonstrating at least some commitment to help with the basic subsistence needs of the household.

The *maja de manzana* has taken on a more significant cultural role in Chiloé over the past several decades. As the region's economy has shifted toward wage labor, islanders in both rural and urban sectors of Chiloé have become increasingly occupied with full-time jobs. This has left little time for mingas, little time for subsistence agriculture, further contributing to the decline of reciprocal-labor traditions, and further accelerating the pull toward a cash-based economy. In most of Chiloé, mingas have become a practice of the past.

The maja, however, has bucked this trend. Because it can be carried out on weekends when neighbors with wage-paying jobs are available to help, it has remained strong, one of the most widely practiced forms of labor reciprocity in Chiloé today. It is seasonal, festive in atmosphere, less time consuming than other mingas, and less labor intensive. All of these factors have contributed to its survival as a cultural practice.

---

[7] Fieldnotes, Llingua, March 2006.

Majas are therefore viewed by many islanders as one of the last remaining mingas. The practice has taken on a meaning that extends beyond its simple function of producing chicha. Majas represent the core values of reciprocity and solidarity that many Chilotes identify with: work carried out not for the sake of earning cash and accumulating profit, but for the sake of strengthening a mutually beneficial long-term relationship. And the drink that is produced from majas—chicha—has, for similar reasons, come to embody particular cultural ideals. As described to me by Renato Cárdenas, chicha often represents "the solidarity of the occasion. Just like *mate* [a regional tea], the act of drinking is a communal one in which the cup is shared. No one is granted special privileges; everyone draws from a common source."[8]

Concluding Remarks

Potatoes, curantos, and chicha are similar in one major respect: they involve reciprocal labor traditions that are at the core of rural Chilote culture. The common denominator between them, in other words, is the minga. Only by gathering neighbors for a work party can potatoes can be harvested, curanto feasts organized, and cider produced. The simple act of receiving help obliges one to return help at a later date, perpetuating indefinitely those work relationships. Historically, this dynamic has applied to nearly all households in Chiloé; today, it is true for rural sectors of the archipelago where cash is still in limited circulation. However, even with the decline of reciprocal labor traditions in recent decades, nearly all islanders today assign special status to the three foods described in this chapter. More than simple pastimes or sources or nourishment, they are emblems of a larger islander identity, symbols of a history of hardtack fishing and farming, of neighborly assistance and solidarity, that is held in high esteem across much of Chiloé today.

References

Ames, M., and D.M. Spooner. 2008. DNA from herbarium specimens settles a controversy about origins of the European potato. *American Journal of Botany* 95 (2): 252–257.
Byron, John. 1812. *The Narrative of the Honorable John Byron*. Edinburgh: James Ballantyne & Company.
Cárdenas Álvarez, Renato, Dante Montiel Vera, and Catherine Grace Hall. 1991. Los Chonos y los Veliche de Chiloé. Santiago de Chile: Ediciones Olimpho.
Cárdenas Álvarez, Renato. 1996. *Chiloé: Diccionario de la lengua y de la cultura*. Santiago de Chile: Ediciones Olimpho.
Cárdenas Álvarez, Renato and Carolina Villagrán Moraga. 2005. Chiloé: Botánica de la cotidianidad. Santiago de Chile: Gráfica LASCAR.

[8] Interview, Calen, Nov.11, 2006.

Cárdenas Álvarez, Renato, Lorna Muñoz Arias. 2015. *Chiloé Contado Desde la Cocina*. Santiago: Talleres de Gráfica LOM.

Castro San Carlos, Amalia. 2016. Chicha y Sidra de Manzana en Chile (1870-1930): Manzanas con Identificación de Origen. *RIVAR* 3(9): 4–25.

Darwin, Charles. 1998. *Chiloé*. Santiago de Chile: Editorial Universitaria.

Daughters, Anton. 2014. Of Chicha, Majas, and Mingas: Hard apple cider in twenty-first century rural southern Chile. In *Alcohol in Latin America: A social and cultural history*. Tucson: University of Arizona Press.

———. 2016a. Fish kills and protests on the islands of Chiloé. *Anthropology News* 57: 61–66.

———. 2016b. Southern Chile's archipelago of Chiloé: Shifting identities in a new economy. *Journal of Latin American and Caribbean Anthropology* 21 (2): 317–335.

Dillehay, Tom D. 1989. *Monte Verde: A late pleistocene settlement in Chile*. Vol. I. Washington and London: Smithsonian Institution Press.

Dillehay, Tom D., C. Ramirez, M. Pino, M.B. Collins, J. Rossen, and J.D. Pino-Navarro. 2008. Monte Verde: Seaweed, food, medicine, and the peopling of South America. *Science* 320 (5877): 784–786.

Draghi Lucero, Juan. 1940. *Fuente Americana de la Historia Argentina. Descripción de la provincia de Cuyo, Cartas de los Jesuitas Mendocinos (1787)*. Mendoza: Best Hermanos.

Galindo Cárcamo, Herman. 2002. *Chiloé: Tierra de Todos*. Santiago de Chile: Impresos Euskadi y Cia Ltda.

Jiménez, Miguel. 2009. *Relatos de mi Niñez y Otros Escritos*. Santiago de Chile: Imprenta El Sur.

Kaspar, K.L., J.S. Park, C.R. Brown, B.D. Mathison, D.A. Navarre, and B.P. Chew. 2010. Pigmented potato consumption alters oxidative stress and inflammatory damage in men. *The Journal of Nutrition* 141 (1): 108–111.

Keynes, Richard. 2000. *Charles Darwin's zoology notes and specimen lists from the HMS beagle*. Cambridge: Cambridge University Press.

Lizárrago, Fray Reginaldo de. 1935. *Descripción Breve de Toda la Tierra del Peru, Tucumán, Rio de la Plata, y Chile, para el Excelentísimo Señor Conde de Lemos y Andrada, Presidente del Consejo Real de Indias. 2ª Edición. Libro Primero*. Buenos Aires: La Facultad.

Nabhan, Gary Paul. 2016. Parque de la Papa. In *Ethnobiology for the Future: Linking Cultural and Ecological Diversity*. Tucson: University of Arizona Press.

Nuñez de Pineda y Bascuñan, Francisco. 1973. Cautiverio Felíz y Razón Individual de las Guerras Dilatadas del Reino de Chile. Santiago de Chile: Editorial Universitaria.

Pardo, Oriana, and José Luis Pizarro. 2005. *La Chicha en el Chile Precolombino*. Santiago de Chile: Editorial Mare Nostrum.

Pérez Rosales, Vicente. 1886. *Recuerdos del Pasado: 1814–1860*. Santiago de Chile: Impresoras Gutenberg.

Pfeiffer, Evelyn. 2016. Chile's Record Toxic Tides May Have Roots in Dirty Fish Farming. *National Geographic*. May 17.

Ristaino, Jean Beagle, and Donald H. Pfister. 2016. 'What a painfully interesting subject:' Charles Darwin's studies of potato late blight. *Bioscience* 66 (12): 1035–1045.

Rivas, Pilar and Carlos E. Ocampo. 2005. El Antiguo Curanto Chilote. *Chile: País Oceánico*. Santiago de Chile: Ocho Libros Editores.

Rosales, Diego de. 1878. *Historia General del Reyno de Chile, Flandes Indiano*. Valparaíso: Imprenta El Mercurio.

———. 1989. *Historia General del Reino de Chile, Flandes Indiano*. Santiago de Chile: Editorial Andrés Bello.

Stuart, James S. 2010. Curanto: Chiloé's Ancient 'Clambcake.' Aug. 2. Retrieved online on 3/6/2018: http://eatingchile.blogspot.com/2010/08/curanto-chiloes-ancient-clambake.html

Thoms, Alston V. 2009. Rocks of ages: Propagation of hot-rock cookery in western North America. *Journal of Archaeological Science* 36 (3): 573–591.

Tornero, Recaredo S. 1872. *Chile Ilustrado*. Valaparaíso: Librerías I Agencias del Mercurio.

Yao, Stephanie. 2010. True origins of widely used potato germplasm revealed. *Agricultural Research* 58 (5): 22–22.

Chapter 5
Trueque Chilote: Traditional Barter Networks Connect Nature and Society in Northern Patagonia

Richard A. Vercoe

> "Thus began a learning that cultivated plants are living artifacts of times past, available where archeology and written document are wanting, or making these more explicit."
> (Carl O. Sauer, Agricultural Origins and Dispersals. 1969)

Introduction

The agrarian system of Chiloé is as unique and dynamic as the archipelago's UNESCO World Heritage recognized culture. This agricultural hearth incorporates both marine and terrestrial systems in a food system that has sustained the islands' inhabitants for thousands of years, while conserving and even maintaining highly diverse ecological systems such as temperate rainforests, brackish estuaries, freshwater wetlands, deep water fjords, and endless coastlines. The result is a resilient and highly sustainable agrobiodiverse landscape that was recently designated as one of the world's first Globally Important Agricultural Heritage Sites by the United Nations' Food and Agriculture Organization (FAO 2011).

The endemic potato-wood barter system between the islands of Chiloé and the surrounding fjordland communities of Patagonia, known locally as *trueque Chilote*, has been in place for centuries.[1] Fjordland communities along the rugged mainland coast exchange wooden posts cut from native trees for the abundant potatoes grown in Chiloé. This barter system provides vital resources to isolated communities where national currency is scarce or nonexistent. The natural and cultural dynamics

[1] The English translation of "Trueque Chilote" is Chilote barter. Chilote barter will be used in the text to refer to the traditional, nonmonetary exchange of local products (specifically potatoes and wood) that is unique to the culture Chilote people of the Chiloé islands and mainland coasts of southern Chile.

R. A. Vercoe (✉)
Department of Geography, University of Wyoming, Laramie, WY, USA
e-mail: rvercoe@uwyo.edu

A. Daughters, A. Pitchon (eds.), *Chiloé*, Ethnobiology,
https://doi.org/10.1007/978-3-319-91983-6_5

of trade based on primary food production and natural resources in these rural communities offers insights into the spatial organization and social continuity that non-monetary barter systems are uniquely able facilitate.

This chapter presents a cultural ecological analysis of Chilote barter. It describes how local agroecological systems can give rise to particular cultural practices, landscapes, and sustainable economies. Based on 2 years of field research and more than a decade of immersion in the region, this study focuses on one of the more unique and previously undocumented components of the Chiloé archipelago heritage agroecological system. I examine how the Chilote barter system is still being practiced within communities of the region, including the specific items being traded, exchange rates, and the way the trade is driven primarily by natural resource needs and sociocultural relations.

For centuries, the residents of Chile's Chiloé archipelago have maintained resilient communities and sustainable landscapes with high crop diversity, strong interdependent community networks, and unique socioecological lifeways. Chiloé is the home of several important global cultivars, most notably the potato *Solanum tuberosum tuberosum* and the strawberry *Fragaria chiloénsis* as well as the only low-elevation adapted quinoa variety *Chenopodium quinoa*. The islanders—called Chilotes—have cultivated hundreds of varieties of potato since before the arrival of the Spanish who subsequently introduced the tuber to Europe. While domestication of the potato family occurred in both the Altiplano region of Bolivia, Peru, and Ecuador and the Chiloé archipelago, it is the high latitude, low elevation, and climactic similarities of Chiloé that facilitated potato exports during the seventeenth and eighteenth centuries. Chiloé's climate and low-lying island biogeography is the southern hemisphere equivalent of the British Isles and more specifically, Ireland. This climactic connectivity accounts for the ease with which the potato became naturalized in the Irish landscape and culture as well as in many other Western European regions. In fact, recent genetic studies confirm that 90% of the genetic material in all of the world's major potato varieties originate from the lush fields along the shores of the Chiloé archipelago rather than the more emblematic terraced slopes of the high, arid Andes (Yao 2010).

Research Methods and Analysis

This study took place on both the mainland and island sides of the trade network. I initiated research visits with known wood-traders who then established and facilitated contact with island potato cultivating counterparts. Subsequent participants were obtained through discussion and interviews with community members who were familiar with the barter activity and knew other local participants on either or both the wood and potato ends of the network. I quickly found that multiple potato-producing contacts could be obtained from a single wood-trader and likewise, multiple new wood-trader interviewees were identified through contact with the potato producers. In this regard, this study was not based on a random survey of residents within the geographic area of study or the trade network. Rather, I used a purposive sampling procedure to select islanders who were known to participate in barter. Because of the remoteness and lack of formal connectivity in this area—the fact that there were little or no formal communication and transport services—this methodological approach was more viable. Of the 18 semi-structured interviews I carried out, 10 were with mainland wood providers and 8 with island potato producers.

Interviews were conducted at the informant's residence in most cases. However, I obtained some follow-up information by participating in actual trading journeys, as well as by working in the forests and fields helping harvest wood or potatoes. The semi-structured interviews provided quantitative data regarding land ownership location and dimensions, productivity rates and regeneration rates, dates of trade voyages and quantities traded, as well as qualitative data describing social relations and barter dynamics. The quantitative information allowed me to establish an ecological baseline of land-use and practices. The interviews gave me ethnographic information to help address the cultural aspects of the nature-culture dynamic of this barter system. Qualitative data from semi-structured interviews includes seasonal patterns of production, preference and selection of specific varieties cultivated, which varieties were for domestic consumption, trading, or both. Additional data included harvest techniques, and relationship dynamics such as cooperative labor practices or selective use of animals such as oxen which were shared between multiple households.

I most often conducted the interviews at the informant's home. This practice was mutually beneficial because, given that transportation is limited or nonexistent, participants did not have to expend time or energy traveling to another location. Few families in this region have personal transportation other than by foot. Conducting interviews "on site" also allowed me to visit the participant's land from which they most often obtained their goods to trade. Since it is local custom to receive guests into one's home for conversation, participants were always amenable to these visits. Due to the Chilote's hospitable customs, interviews usually took place around a wood-burning stove with numerous rounds of mâté (a green tea that is traditionally consumed during social interactions). Other family members were often present during the interviews, since it is traditional practice within the culture to live with extended family and most daily activity revolves around the homestead. Interviews intentionally occurred around the participant's main meal, lunch, when they were more likely to be home and available. This occasionally led to interviews lasting two or more hours with an entire family and a meal included. This allowed for several digressions from the interview questions, which provided interesting and valuable context beyond the formal questions.

A cultural ecological land-use analysis of the area included data obtained from the Military Geographic Institute of Chile (IGM), preclassified land-cover layers obtained in the field, and researcher generated land-cover classifications based on satellite imagery.

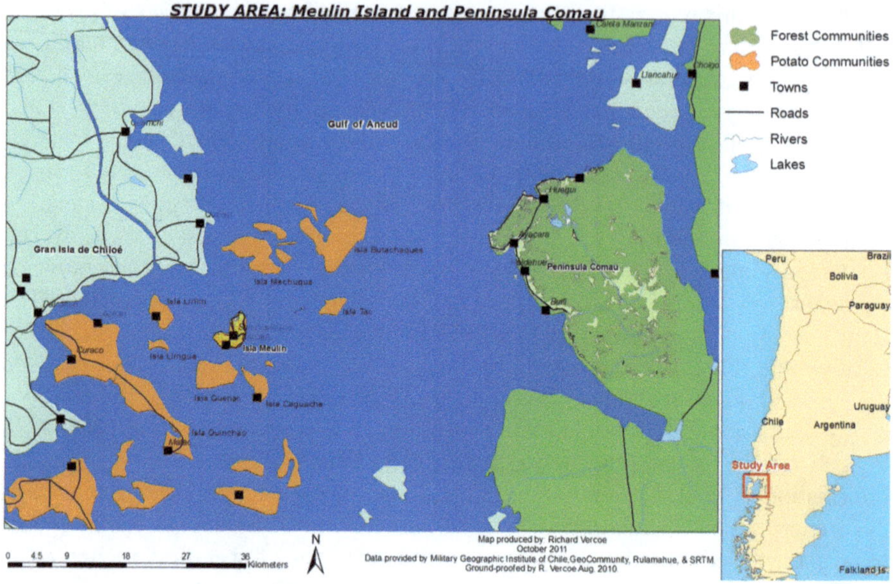

Fig. 5.1 Map of study area. (Map credit: Richard Vercoe)

Description of the Study and Area

The study covers 360 km² from the Comau Peninsula along the Andean mainland and west across the Sea of Chiloé to Meulín Island. This area is in the Tenth Region of the Republic of Chile, between 41 and 44 degrees South latitude and 72–74 degrees West longitude (Fig. 5.1).

Meulín Island is one of the many interior islands of the Chiloé archipelago. This network of islands has very limited relief (500 m maximum) and sandy loam soils, ideal for root crop production. The continental Chiloé area, as part of the Andean Orogeny, has a much more pronounced relief (up to 2500 m within a kilometer from the coast) marked by glacier-capped peaks along the eastern boundary. The soil composition of the inhabited lowland areas of the peninsula is characterized by boggy soils that contain much less sand and greater amounts of glacial till and gravels on igneous bedrock, resulting in very poor soils for crop production of any kind. The island areas receive about 1000–3000 mm of annual rainfall whereas the continental areas may receive several times as much precipitation, from 4000 to 8000 mm, due to the orographic effect of the Andean uplift. This high differentiation in precipitation between the islands and the mainland results in the islands maintaining an ideal balance of year-round growing conditions while the mainland area literally becomes saturated throughout the year resulting in soils that are flushed of their crop nourishing nutrients.

The Potato People

Meulín is one of dozens of interior islands scattered throughout the Gulf of Ancud between the big island of Chiloé and the mountainous fjordlands of Continental Chiloé. The island's low profile, rolling terrain, rich soils, and proximity to Chiloé's rainshadow protect it from the prevailing storms and strong seas arriving from the open Pacific Ocean to the west. This orientation provides Meulín with the optimum climatic and ecological conditions for potato cultivation. Its approximately 2000 inhabitants are dispersed fairly evenly over the island's 13 km^2. Two official communities (San Francisco and El Transito) each serve the rural population with limited services such as a school, church, and a pier as well as a few general stores operated as extensions to family homes. These stores offer basic goods such as cleaning products, food staples, and canned items, as well as, on occasion, fresh fruits and vegetables, depending on the season. All of these items are "imported" from urban areas to the north and have an elevated cost and due to limited transportation options in this remote region. The island's communities have a long history of potato production that is a vital sustaining component of their traditional horticultural and fishing lifestyle. Most of these long-inhabited islands of horticultural/pastoral legacy have been largely cleared of the dense temperate rainforests emblematic of northern Patagonia.

Local farmers such as Doña Luz and Don Pato typically trace their family lineage and land holdings back 300 years or so. Potato cultivation and animal husbandry are the primary horticultural activities of residents, while shellfish collecting and hand-line fishing are the primary marine activities. Due to diminished availability of mature trees on many of the smaller islands, potato-cultivators often struggle to separate their livestock from their farm fields. This often results in damage or losses to a percentage of their crops. Therefore, the demand for wood for fencing material is the most important factor driving Meulín residents to participate in the Chilote barter.

All of the interviewees on Meulín Island confirmed that the Chilote barter is currently active in their communities. The property sizes of the island interviewees ranged from 2 to 25 hectares in area with 0.25–1.5 hectares, respectively, dedicated to potato production and the remainder left to pasture (primarily sheep). The Meulín islanders are horticulturalists and pastoralists, exclusively working small-scale plots by hand without any mechanical or motorized tools. Those who were interviewed only used, altogether, 8% of their land for potato production at any given time; however, I found that, overall, 54% of the total land on the island is used for some type of agricultural production; 27% consists of shrubs or immature forest, 11% wetland, and 9% bare soil (the latter occurring primarily along the intertidal zone). The high percentage of agricultural land is largely due to the fact that the domestic animals (mostly sheep) require extensive pasture for grazing. Animal husbandry is not only the islander's primary source of protein but also provides limited cash income when animals are taken to the big island or mainland for sale (Fig. 5.2).

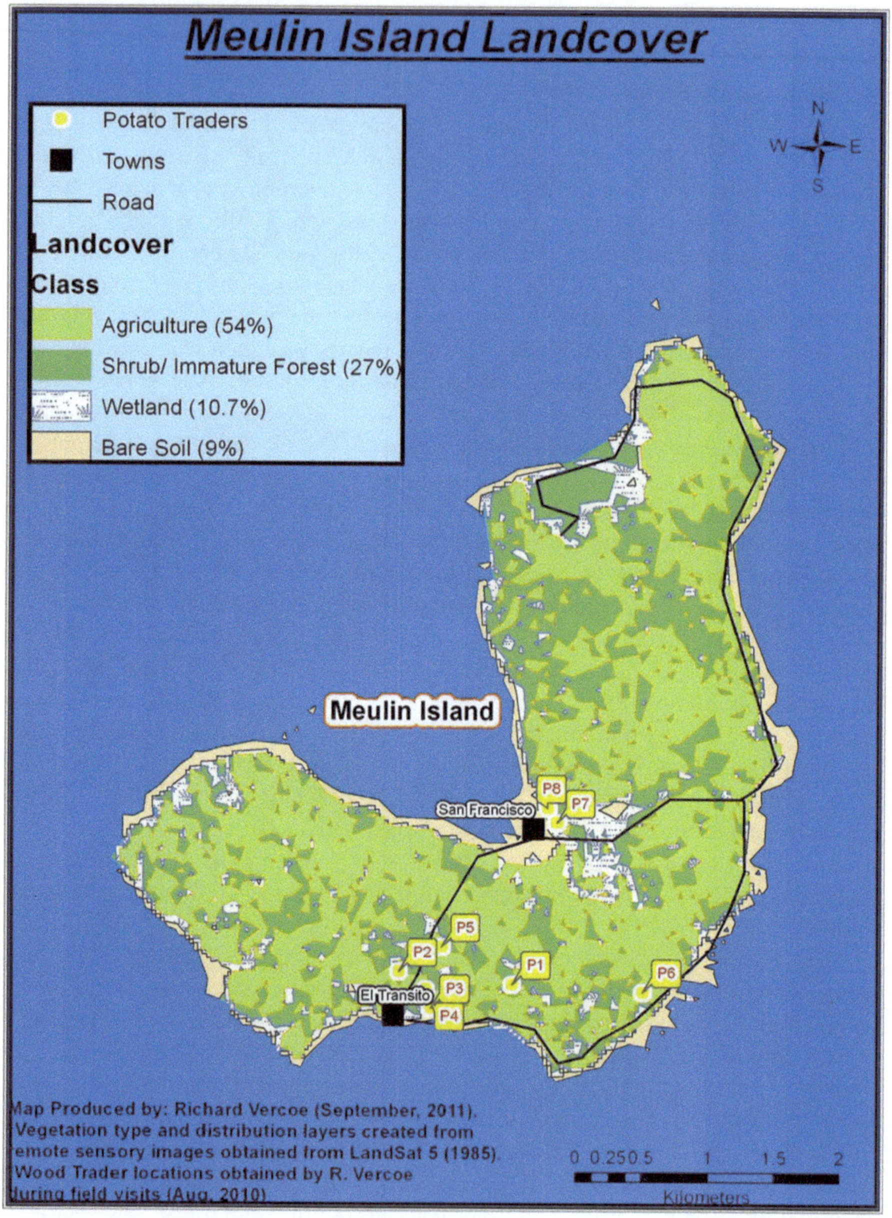

Fig. 5.2 Map of Meulín. (Map credit: Richard Vercoe)

The islanders keep sheep, chickens, and pigs, with sheep being the most common domestic animal, kept by all of the interviewees. Ten sheep required approximately one hectare without additional sources of feed such as hay, grains, or potatoes, according to Don Pato. Pigs and chickens were the next most common type of livestock. These were penned in the evenings and during certain planting or harvest times, but otherwise left to roam within the property along with the sheep. The chickens and pigs were allowed to forage on the property as well as being fed discarded and leftover vegetable waste from the homestead. This results in a highly sustainable land management system that rarely requires external material inputs other than an occasional marine algae harvest along the shoreline to replenish nitrogen and phosphorous levels in the potato fields. The sheep not only provide meat but also wool for various domestic uses as well as rich guano to add to the fields as fertilizer. All of the interviewees that raise sheep said that they derived a limited cash income from the seasonal sale of some of their sheep to neighboring islanders and people from the mainland. This was generally the only, or at least the largest, cash income source on an annual basis for the islanders interviewed. This cash was then used to purchase essential items such as barbwire, tools, home furnishings, and other domestic materials that cannot be obtained by barter.

Potatoes in Their Environment

Meulín potato-cultivators achieve fairly high productivity by sharing labor among neighbors and using management practices based on traditional ecological knowledge developed over hundreds of years. The soil quality and temperate climate on the island provide an ideal foundation for crop success. Indeed, soil and climactic conditions throughout Chiloé are dramatically different from those of the rocky, nutrient leached, and boggy rainforests of the mainland peninsula. Productivity rates for potatoes are approximately 7500 kilos per hectare and were determined using estimates from several interviewees. For example, Don Pato's most recent harvest yielded seventy 50 kg sacks of potatoes from a half hectare parcel. The 50 kg sack is the standard unit of measure in the region for storage, transportation, and trade.

The island's sandy loam soils are well-suited for growing crops such as tubers. The soils maintain their nutrient load well and receive an annual rainfall of 2000–4000 mm, which is half that of the cloud-wringing, mountainous peninsula just 80 km to the east. The finer, more porous and even textured soil of Meulín also facilitates cultivation by hand tools, such as digging-sticks and long-handled hoes. Sandy loam soil can support a rich nutrient base. Potato-cultivators maintain soil health not only with sheep guano but also by turning-in marine alga when preparing the soil to further supplement the nutrient base. All of the interviewees rotated their potato fields either every year or two to reduce the permanence of soil-based plant pathogens. No irrigation, chemical fertilizers, or pesticides were used by any of the families that I interviewed.

There are at least 12 native potato varieties currently propagated on the island. Most can be found in each of the farmed fields any given year, but proportions of each vary between farmers. The exact number of total potato varieties on the island is unknown, but Don Cloro identified a dozen by name during the interview, and

Image 5.1 Potatoes of Chiloé. (Photo credit: Richard Vercoe)

several other respondents confirmed that number. Currently the *pie amarilla* (yellow foot) and the *mechuña* (purple stain) varieties are the preferred ones for cultivation based on their taste, resistance to disease, and storage longevity. All of them are considered native to the area, and there are currently no industrial potato varieties cultivated by any of the interviewees. In fact, all of the interviewees had tried industrial varieties in the past for one reason or another and unanimously preferred the local varieties for taste and texture (Image 5.1).

The farmers articulated two reasons for preferring native varieties to industrial ones. The first was that industrial varieties were much more susceptible to disease, rot, and more sensitive to varied conditions of water, soil health, frost, or sunlight. Farmers saw them as "higher maintenance" than their locally adapted varieties. They also stated that industrial varieties were dependent on the chemical fertilizers that were originally provided to the farmers when they were first introduced as part of state-sponsored agricultural and rural development programs. However, once the programs ceased providing the chemical fertilizers, the farmers found it difficult to maintain the crops with their traditional resources and methods.

The second reason was the poor taste and consistency of the industrial varieties. The islanders described them as being tasteless and having a gritty, grainy, or mealy consistency that often resulted in them falling apart when cooked. This perception was corroborated by the mainland traders who refused to accept the industrial varieties in exchange for their wood. The selective power of taste and preference for the native varieties over the industrials is a testament to the influence that the regional barter system has, in this case an influence that supersedes that of the national currency and marketplace. If not for the relative isolation of this region from commercial markets, the locally less desirable industrial varieties might have replaced the native ones as an export crop.

Días Cambiados

Both the wood-traders and the potato producers identified a cultural practice called *días cambiado* (day exchanged) as a fundamental part of the mutual aid they rely on to help cultivate and/or harvest the products they bring to the barter. It is common practice on the peninsula for one neighbor to help another harvest wood, gather a load sufficient for a trip, or transport another's wood for trade. On the islands, shared days preparing fields, planting, and especially harvesting the potatoes are all common labor-exchange activities by the potato producers. According to Don Pato, the *días cambiado* tradition has a long history in Chiloé and is essential to the success of his community as a whole. It enables each family or individual the opportunity to maximize productivity and maintain social connectivity. These two points were corroborated by all of the other interviewees.

A *días cambiado* is essentially a labor barter between neighbors, a type of *minga*. The exchange is usually an exchange of equal services for equal amounts of time. This labor was performed by men and women (70% men and 30% women of those surveyed) in the case of the potato-cultivators. According to interviewees, the most common day exchange occurs during potato planting and harvesting season. Several neighbors convene at one property to help prepare soil, plant, harvest, or carry a harvest to storage. The host is expected to serve visiting day exchangers three meals and several maté tea breaks and often a bit of homemade hard cider (*chicha*) at the end of the day. The meals and tea breaks are ripe opportunities for dialogue over topics and activities pertinent to the community's livelihood. The host will reciprocate the labor for every person that worked on his/her behalf if possible. Doña Luz shared that the social support and interaction from this type of minga is even more important for her now that she is a widow who tends her land alone. She is not able to perform the heavy lifting required to harvest her potatoes, provide for her two pigs, and trade for fencing and firewood.

The interview results from my research suggest that the *días cambiado* appears to be practiced more frequently and has great social significance among the island farmers than more remote mainland wood-traders. This may largely be due to the greater density and accessibility of the island communities, as well as the nature of the more predictable seasonality of the potato harvest. Beyond the peculiarities of each activity, the *días cambiado* both increases individual productivity, whether for potato-cultivators or wood-traders, and provides vital social connectivity in rural and isolated landscapes.

People of the Forests

The 612 km^2 Comau Peninsula sits at the base of the Andes Mountains and forms the mainland boundary of the Sea of Chiloé, approximately 80 km east of Meulín Island. This region contains some of the largest swaths of virgin coastal temperate rainforest remaining in the world (Nahuelhual et al. 2007). The Comau Peninsula has experienced more recent habitation than the islands. Living within the confines of a very formidable and largely intact virgin temperate rainforest, these remote peninsula communities are more culturally and economically isolated than most communities on the Chiloé archipelago. The inhabitants of the peninsula rely on the dense coastal temperate rainforests that cling to the steep coastline, in addition to fishing, to provide for their subsistence lifestyle. All of the approximately 2200 inhabitants live within half a kilometer of the coast, leaving the rugged, rainforest interior almost entirely untouched (Fig. 5.3).

Analysis of LandSat remote-sensing imagery determined that 85% of the 612 km^2 Comau peninsula is covered with native forest. This value was corroborated by anecdotal descriptions from the interviewees. The land-cover data identified an additional 14% of the total area as shrubs. The ethnographic and field data determined that this GIS layer was misidentified. All of interviewee sites were located at the outer edge of areas identified by this classification. The interview information was combined with site visits that included GPS referenced photographic records. The integrated data identified much of the coastal shrub area to actually be cleared-forest used as sheep pasture and general living space surrounding homesteads. The ethnographic ground-proofing of the land-cover data reduced the total shrub coverage by half resulting in 7% shrub coverage and 7% agricultural land. The analysis concluded that only 7% of landscape on the peninsula is significantly altered due to human activities.

There are four communities on the Comau Peninsula: Ayacara, Buill, Huequi, and Poyo. These communities are linked by a single dirt road that provides access to basic services for the rural households spread along the shoreline. The difficulty of producing a stable food source on the Comau Peninsula is the most important ecological factor driving the participation of the wood-traders in the Chilote barter. The rain drenched soils found on the peninsula are generally flushed of nutrients and saturated with water much of the year, creating bog conditions that are neither productive for crops nor tolerant of grazing pressures. Several interviewees identified poor soil conditions and high precipitation (up to 8 m per year), due largely to the moisture wringing effects of the nearby Andes, as reasons for limited production capability on the peninsula.

All of the wood producers interviewed explained they had a small potato patch on their land, but that it was never productive enough to sufficiently provide for the family and was often prone to failure due to soil saturation, poor nutrient conditions, as well as mold and fungus outbreaks. Locals such as Don Oscar and Don Edelmo described the potato fields of the island potato producers to be less humid, have a more sandy loam texture, and be more productive than any of the peninsula's soils. Woodland locals describe the potato-producing islands in almost mythical terms, as a land where an endless variety of tasty potatoes grew effortlessly among rolling pastures containing well-fed sheep grazing on lush grasses, but not a tree to be found. The preference for the taste of the local Chilote potato varieties and the ability to obtain the staple food item without cash are the primary cultural and economic factors driving the wood-traders' participation in the Chilote barter.

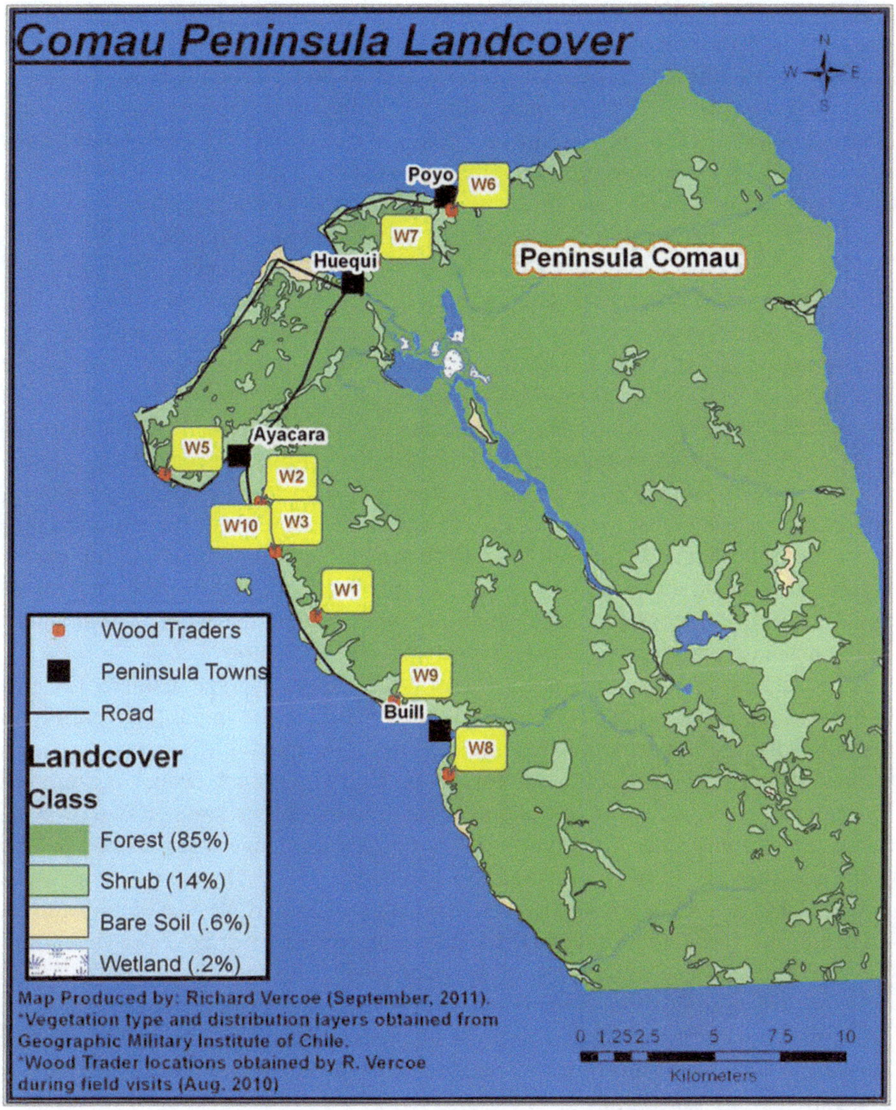

Fig. 5.3 Map of Comau Peninsula. (Map credit: Richard Vercoe)

Chilote barterers can be found in all of the communities on the Comau Peninsula. Four of the ten wood-traders interviewed came from the main town of Ayacara. In addition to the information they provided, they helped me establish other wood trading contacts in the neighboring coastal towns of the peninsula. The most common item that was sourced for barter by the woodland communities was wood for fence posts. Each post is made from a tree trunk 10–15 cm in diameter cut into 2 m lengths.

Common harvesting and removal techniques consist of axe felling and shoulder-carrying, with the occasional use of oxen. Only one participant used a chainsaw to cut the harvested trunks to post length prior to loading them on the boat. All others used an ax for all of the wood processing, from felling to limbing and trimming. Chainsaws and the fuel they require are prohibitively expensive for this largely non-monetary lifestyle and thus more of a luxury in the region. Don Gabo, Don Samuel, and Don Chochi all made use of oxen to drag bundles of ten to twenty 10 m long trunks at a time, but it was most common for harvesters to shoulder-carry the trunks one or two at a time, using only a thick wool sweater as padding.

The most common tree species harvested for posts are the luma (*Amomyrtus luma*) and tepú (*Tepualia stipularis*). Luma is from the myrtle family and can grow to 20 m in height. It is favored by wood-traders for its multiple trunk growth form and quick asexual regeneration. This dominant understory tree species favors high regeneration rates as well as year-round harvestability. The luma is also one of the more rot-resistant species found in this humid climate. Luma fence posts are alleged to last between 10 and 20 years, depending on the location and environment in which the post is placed. Even when cut and limbed, fresh-cut fence posts commonly exhibit a limited regrowth if set in fertile bare ground. The ability to regrow may further extend the wood's longevity for up to a decade. Luma grows in thickets and has thin flexible branches with small evergreen leaves. The wood type and flexibility make it convenient to harvest by axe and shoulder-carry.

Tepú, also a member of the myrtle family, has a larger average diameter (20 cm) but lower average height (10–15 m) than the luma. Like the luma, the tepú has an abundant distribution and quick regeneration rate. One major difference is the tree's longevity. One post can last 20–30 years in the ground. Luma, however, is preferred by both the wood- and potato-traders. Wood-traders like its smaller size, ease of harvest, and transportability. Potato-traders find the exchange rate for luma preferable since they get more posts per sack of potatoes.

Luma and tepú trees are said to regrow equally well in disturbed and undisturbed climax forest. Don Samuel explained that in order to avoid overharvesting a particular area, his family selectively cut different areas on a 10-year cyclical rotation. He said that this practice had been handed down over the generations, taught from father to son. Don Samuel was actively teaching his son and grandson this selective harvesting technique in order to maintain a sustainable practice for future generations.

I found the average property size of participants was 56 hectares, with a high of 120 ha and low of 20 ha. Of this area approximately 80–95% was forest and the remainder was cleared land for the home, some sheep grazing and a small vegetable garden. Don Chochi maintained 30 ha out of 120 hectares cleared for grazing cattle and sheep. He was the only interviewee that raised cattle on the peninsula.

The Making of Forest Waterways

One day Don Samuel invited me along for a full day of luma harvesting in preparation for a barter voyage. The family maintains a pair of oxen for load-bearing tasks around their property. They use it, for example, to transport items to and from their boat that is anchored about a kilometer from the house. I accompanied Don Samuel and his two sons. One son—the captain of the family boat—was in his mid-twenties, the other in his early teens. We spent several hours at the beginning of the day trying to locate the oxen that had been turned loose into the forest to forage. The coastal temperate forest of that sector is dense from ground to canopy, with many intermediate layers of native bamboo stands, the multiple-trunked luma, and tepú mid-canopies, all within the canopies of the larger birch and beech species. Moving through the understory and on top of boggy and muddy ground is challenging, especially when trying to carry an ax or haul 10 m long trunks. The forest is even more impenetrable when trying to move two yoked oxen pulling 20 or 30 of these trunks.

Don Samuel's family practices a traditional Chilote land management technique. They utilize their land's natural features and topography to develop small waterways by which to transport harvested wood. Pulled by the oxen, the wood partially floats down cobbled, clear streams. Year after year, as the family moves around the forest, they travel along small drainages in the undulating landscape and remove dense shrub layers. Because of this movement, peat moss and small grasses are eroded by heavy rain, run-off, and disturbance, thus creating a specific network of small stream channels of approximately 10–20 cm deep (depending on the rain that day), about 2 m wide (the width of an oxen team), eroded down to the cobbled glacial till and bare rock substrate rather than the knee deep mud and peat bog.

In order to harvest, the family, with its oxen, walks up a waterway until a desired location is found. Family members fan out into the forest to cut and limb tree trunks and then carry them by hand back to the oxen waiting in the waterway. When about 20–30 limbed posts are harvested, they are stacked together and chained onto the yoke between the oxen, and the team half-drags/half-floats the bundle out of the forest toward the house. The entire harvest outing, from finding oxen to harvesting and returning home to buck trunks into posts, takes 3 men approximately 8 h and yields 200 posts (Image 5.2).

Analysis of the remote-sensing imagery was unable to pick up any signs of these manmade waterways. This was expected considering the satellite imagery's 30 m resolution combined with the area's dense canopy coverage. This is where direct field documentation served well, not just for the value of ethnographic data but also for documenting a little-known, traditional land management technique that has a localized ecological impact on water quality and distribution, vegetation, and soil erosion. However, the waterways follow natural drainage courses in low relief areas and do not appear to erode any further than the initial 30 cm of topsoil down to the cobbled and hardrock substrate. The interviewees claim that the waterways reduce overall impact on the forest by channeling foot and oxen traffic along these narrow, nonvegetated corridors rather than typical forest travel that requires cutting and

Image 5.2 Transporting posts down waterways. (Photo credit: Richard Vercoe)

breaking of vegetation and branches along constantly varying travel routes due to the dense vegetation.

The Trading Voyages

Trading voyages can be quick and productive, or they can be fruitless feats of dash and daring across rough seas with not a single trade made. Whatever the results, the goal is the same: trading wood for potatoes and other products. I accompanied Don Samuel and his two sons on a trading voyage in early August (mid-winter). We had trouble finding people with potatoes available to trade, and we had to navigate through a Patagonian storm so fierce that literally all of the ports in the region were closed by mandate of the Chilean Navy. Long days and nights were spent keeping the small cargo boat afloat with its load of potatoes and wood. For several days the four of us slept, cooked, and maneuvered the boat in rough waters, driving rain, and biting winds. Daily forays ashore in search of islanders with potatoes to trade yielded many stories and granted us insights into the life experiences of Chilotes living among the seas, storms, and mountains of Patagonia (Fig. 5.4).

My research found that all trading voyages are initiated exclusively from the wood-provider side of the network. Ecological conditions such as the peninsula's low crop productivity encourage the barter. Raw materials for boat construction

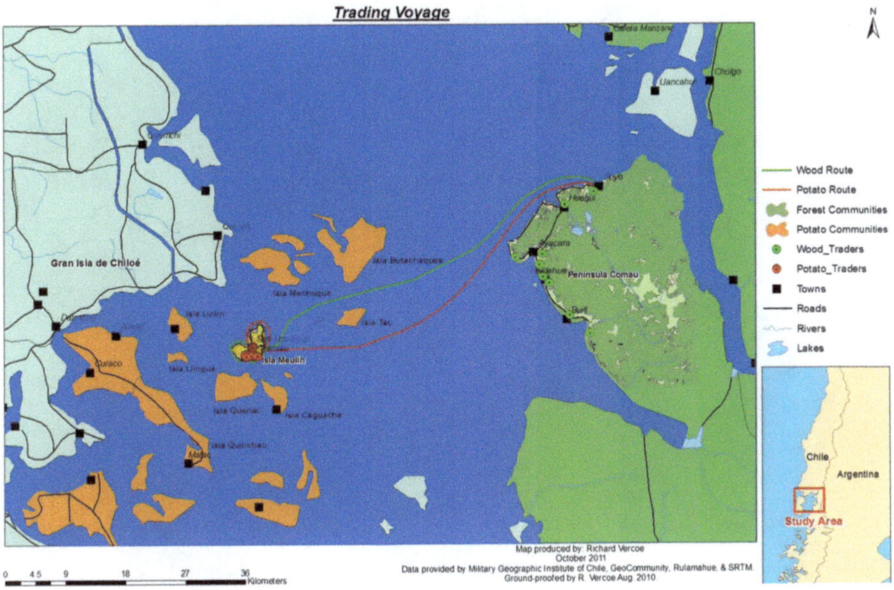

Fig. 5.4 Map of trading voyage. (Map credit: Richard Vercoe)

only exist on the peninsula, thus voyages are initiated from the side that has the better ability to travel. A wood-trader may combine resources with a neighbor in order to fill a boatload of material if they are unable to obtain a full load on their own. Once the potatoes are traded for, the boat owner will "charge" a prearranged amount of potatoes as a handling fee and pay the remainder in potatoes to the neighbor. All wood-traders confirmed a preference for sharing resources and collectively harvesting enough wood to fill a boat. Don Edelmo and Don Samuel, however, preferred to harvest and prepare wood within their family unit. These families had sufficient resources and labor power—a father and several sons—to complete the work on their own, thus simplifying the trade.

Chilote boats are hand-made, hewn by the owners, and constructed from several species of local wood found mostly in mature forests on the peninsula. Traders interviewed for this research used boats ranging from 8 to 12 m in length and 3–4 m wide. The design of a traditional Chilote boat requires large mature timbers. Each boat owner builds their vessel primarily with trees felled on their land. Neighbors help with the labor and by occasionally providing some of the wood if necessary. Labor assistance from a neighbor is treated like the *días cambiados* arrangement practiced on the islands (Image 5.3).

Boats today are powered by small marine motors rather than the sails and oars of earlier times. Trading voyages maximize cargo loads and time commitments in order to justify fuel costs. Smaller boats, such as Don Eliodoro's launch (8 m × 2.8 m) had a capacity of 150–200 luma posts (10 cm × 2 m each), and on return could fill its hold with twenty 50 kg sacks of potatoes (1 metric ton). Some of the larger boats

Image 5.3 Hand hewn boat used for fishing and trade. (Photo credit: Richard Vercoe)

had a capacity of 700–800 posts or approximately 4 metric tons of potatoes. Don Samuel's cargo/fishing boat held 500 posts on our trip.

Wood-trading boat owners are also registered fishermen. Their catches serve as a vital source of protein for the peninsula forest communities where agricultural production is severely limited by environmental conditions. Fishermen recognized by the state as artisanal are allowed to fish for commercial sale and are limited by a monthly quota established by state agencies and apportioned by local fishing unions. The typical monthly commercial returns generate just enough income to cover fuel costs, with some cash leftover for additional domestic purchases. Given that they have a lower agricultural workload as compared to the islanders, the forest communities dedicate more time to the construction, maintenance, and operation of their boats, and to fishing activities.

According to all of the wood-traders interviewed, the winter months of June, July, and August were typically when they embarked on trading voyages. This coincides with seasonal storage of potatoes after the fall harvests when supplies are most abundant and farmers have already made allocations for next season's planting and their own subsistence needs. The short days and rough sea conditions of winter lead to decreased fishing activity, thus time available for trade. Don Rene and Don Vitoy both indicated that that time of the year was ideal for potato producers because they were not yet preparing their fields for the following season's crop.

Wood-traders said they typically choose to initiate a voyage once a boatload of posts has been prepared and the weather is conducive. Only one trader, young Rigoberto, indicated he had prearranged his voyage via mobile phone. The others made the voyage without prior notification to the potato islands. All traders sailed to the islands for barter once each winter to acquire sufficient potato stores. Only 30%

of those interviewed made more than one trip—between two and four per year—in which loads were shared with neighbors or extended family.

The New Year was another time specifically mentioned as an active trading season. That's when a limited harvest of "new potatoes"—small, delicate, early-maturing varieties that are planted earlier than the main crop—become available. These delicacies are a favorite item for early summer Christmas and New Year's festivities in the region. Twenty percent of the wood-traders make these special voyages with high-quality posts (straight and thick) to trade for limited sacks of new potatoes and perhaps some hard apple cider. Unfortunately, wood-traders sometimes only return with dregs of last year's potatoes.

The maritime voyages are often the most memorable and venturesome parts of the trading process. A typical voyage takes at least 3 days, maybe longer, depending on two factors: the ability of the wood-trader to find potato-cultivators looking for products, and the weather. Depending on the wind, waves, and currents, the sea crossing can take from 6 to 12 h. Speed and stability are primary factors limiting the geographic range of the trade network for these hand-hewn, small-motor boats with limited fuel-carrying capacity traversing rough Patagonian inner seas. Sailors prefer to travel within a day's journey of their homes for safety reasons and because of fuel capacity limitations.

Two older traders in their seventies and eighties recounted the time it would take them to sail across the seas when they were young. On a good day, they said, they could match the speed of today's motor boats. But the sail was always at the whim of the wind. Nowadays, most boat operators all carry marine band radios for emergency communication (however, none of the 10 boat owners interviewed had GPS navigation systems and only half had marine charts for navigation). Navigation in this region is done either by sight or compass bearing and time.

Trading a full boatload of wood for potatoes in a single day is often possible. When that happens, wood-traders can again set return sail early on the third day. If barterers have difficulty finding trading partners, they might have to sail to several different parts of one or more islands, requiring additional time and resources. Depending on the design and capacity of the boat, the traders may sleep onboard, anchored just offshore in a protected port or bay, or they might stay in the home of one of the potato-producing families they know. During the winter, it is common for strong storms to form quickly, forcing small and underpowered boats to seek shelter in protected coves, sometimes for several days at a time.

Barter

Barter is a transformative act that moves objects, which are different in kind, between "regimes of value" sustained by free and equal actors (Humphery and Hugh-Jones 1992). Direct barter is an exchange network that does not incorporate symbolic valuations (currency) or indirect transactions (credit) that could lead to power inequalities. The typical resource transactions occurring in barter systems have rates of exchange that are mutually agreed upon by the participating parties rather than commoditized, abstracted, and nonnegotiable values. The barter exchange of goods based on equal participation results in minimal cultural or personal transaction costs and is reinforced by the possibility of repeated future transactions resulting in trustworthy and consistent trade relationships (Cellarius 2000).

The limited "market" and intimate nature of participant relations in the Chilote barter network reduces resource exploitation pressures than that of a more commercialized trade where individual responsibility to maintain one's own land is diminished. Each of these traders only has the resources available to them that their land can provide. In the case of the Chilote culture, there is great importance placed on practices that keep the land in maximum health for future generations and to not overexploit resources for short-term gains. Additionally, there is no real incentive to accumulate either the potatoes or the wood since their respective value beyond personal consumption is only in trade for the other as needed. The direct exchange and shared valuation of goods also serves a self-limiting function with regards to the spatial extent of this trading network. The participants in this trading system have a specialized knowledge and utility for the resources. Therefore, trading networks such as the Chiloe potato-wood barter are naturally limited to either local or regional scales in which both products are produced and able to be exchanged with minimum transportation costs. This, in turn, dissuades overextraction due to travel time, the environmental risk of the often tempestuous Patagonian inner seaways, and the relatively high fuel costs in this remote region.

The Chilote Barter

Three core components were identified as a result of this Chilote barter research: set social behaviors, primary barter items with specific exchange rates, and the ability to accommodate unique or specialized trades all without the use of currency. Both wood-providing mainland traders and the potato-cultivating islanders identified these as commonly shared characteristics of the Chilote barter process. As we have seen in the discussion of the voyages, the wood producers are predominantly the initiators of barter by sailing to the islands with a boat full of posts. Rigoberto and Don Samuel were the only two wood-traders that sometimes used cell phones to contact island traders prior to departure. Don Samuel said that it was important for him because he would sometimes build and trade special wood items—animal feeding troughs and wooden plow blades, for example—upon request by his regular contacts on the islands. He said that prior to cell phones, they would take orders for special items from one journey to the next, which meant waiting many months or even a year before the wood-trader might return with the item. All of the other wood-traders stated that they loaded their boats with what they thought were most generally needed items (fence posts and firewood), sometimes having specific quantities already allocated from requests made during the prior trip.

Upon arriving to the coast of an island, the wood-traders would anchor far enough offshore to avoid getting beached during the high daily tidal ranges of this region. Someone would stay onboard at all times to maneuver the boat—either for safety reasons or to sail it around to a property once a barter was arranged by the landing party. The Chilote barter social dynamic became apparent as soon as the mariners made for shore. Usually, the elder family member would lead the landing party. They would begin by visiting known acquaintances to inquire about trades. Each visit began with an extended conversation with the head of the house, typically inside the home of the islander, seated around a table sharing a cup of mâté. Gender did not appear to be a significant factor in establishing a potential trading partner. More important was whether the person engaging in the conversation had decision-making power as a head of the household. Two of the five observed trades involved women heads of house. Casual conversation could last up to an hour before any discussion of actual trading would begin. This format was repeated in all five of the observed trading meetings that made it past the initial point of establishing mutual interest.

Eventually, the conversation would shift to the actual trade and was initiated by either the wood or potato provider. If the potato provider did not need any wood or did not have any potatoes or secondary commodity to trade, the wood-trader would inquire about whether the islander knew of other individuals who might be interested and able to trade. An unsuccessful trading visit could take between 30 min and an hour and a successful negotiation between two interested parties could last several hours and include a meal together at the islander's home. Neither the wood-traders nor the potato-cultivators identified any family or marriage relations that determined their trading partners. Both sides had one or two preferred trading partners that they felt provided a good product, but they were not averse to trading with new contacts offering good quality items to exchange.

Core items of the Chilote barter have specific long-standing exchange rates that are characteristic and specific to this network. The exchange rates are mutually understood baselines from which all trading negotiations begin. The most common exchange rate used is ten luma posts (10 cm × 2 m) for one sack of potatoes (50 kg). Tepú posts fetch a rate of eight posts/sack due to their larger diameter and greater durability. Wood-traders often bring curved, twisted, and split posts to trade as firewood, which has an exchange rate of 1 m^3 for a sack of potatoes. The posts are evaluated by the potato farmers for straightness and the presence of any splitting. These two qualities are primary determinants by the potato-trader as to the final rate of exchange to be accepted. A straight post is preferred for driving into the ground by hand and for securely attaching wire or cross-poles. Split posts are likely to rot sooner than intact trunks with their protective bark in place. Splitting also indicates that the post was not cut recently and will likely not regrow once placed in the ground therefore reducing its potential longevity as a "living" fence. Noticeably curved or split posts will generally require a two-for-one value in counting toward the 10 posts/sack baseline. This plasticity of adapting to specific characteristics in the goods exchanged is never overtly expressed but is quickly introduced when necessary. The general custom is for an interested potato-trader to head out to the beach or onto the boat to individually choose and select the posts before bringing the wood-trader back to the storage shed where the potatoes are stored. This provides a point of entry for one of the two participants to begin assessing the quality of the goods and begin altering the exchange rate if needed.

With respect to the wood-trader's priorities, the potatoes must be free of bruising and any visible rot, mold, or fungus. It is customary for the wood-trader to personally load the 50 kg sacks as they select the potatoes from a central pile. Once the traders have made their selections and agreed on the rate, they transport the items, often with the help of neighbors and the crew members of the boat. This act of respect and mutual aid occurred for each of the successful trades observed.

Potato-cultivators try to produce enough annually to meet four basic needs: to have sufficient harvest to feed the family until the next, to have seed potatoes for the next planting, to supplement animal feed (usually pigs), and to have 5–20 sacks for trade for fence post and firewood. Doña Erika allocated her harvest of 80 sacks produced on half a hectare for a family of five in the following manner: 50 sacks for family consumption, 10 sacks for seed, 10 sacks for pig feed, and 10 sacks for trading. The traditional way to ensure a minimum harvest needs is to plant a diversity of varieties to account for variations in climate or soil and pest conditions from year to year. I found that the Meulín Island farmers cultivated six to ten different potato varieties on average. The difference in which varieties each farmer chose depended on the particular environmental conditions of their fields as well as personal taste. This dynamic of social and ecological variables used for determining the varieties and quantities of potatoes produced by each farmer is an important contributor to the high agrobiodiversity for which this region is known for.

The third characteristic of the Chilote barter is the ability to accommodate unique or specialized trades all without the use of currency. If the islander does not have any potatoes to barter, but is in need of posts, he or she may offer some alternative items that may be of interest to the wood-trader. Livestock is usually the default alternative offered by an islander. A 20 kg lamb will likely get 20 luma posts or a 15 kg piglet could fetch 30 luma posts. I witnessed a particular barter in which Doña Erika had too limited a store of potatoes to trade. While passing through the front pasture on the way out Don Samuel inquired about a piglet and was shortly on his way back to his boat with a bound piglet in exchange for 20 luma posts and 10 tepú posts. All of the farmer interviewees also planted a limited number of seasonal vegetables in gardens near the home. The garden items included carrots, beets, cabbage, and native elephant garlic. If they had a surplus of these items at the time that wood-traders visited they would sometimes include these items for trade.

These alternative trades are much more closely tied to the specific needs and interests of the two parties. A young piglet will require many months or years of care and feeding before it will be worth its weight in meat, whereas potatoes store well and require no further resources or labor. During the New Year's voyages when new potatoes are limited, wood-traders are often eager to trade their finest posts for a few liters of an islander's specialty hard cider (*chicha*) made from low-growing, but very hardy crab-apples. Ten luma posts will fetch 10–15 liters of 6–10% alcohol cider. Homemade hard cider is a highly cherished holiday beverage throughout the region and stores well.

The Chilote barter facilitates the exchange of vital resources while maintaining a flexibility for alternative goods to be introduced when necessary. The different ecological limitations imposed upon each side of the barter network creates a dynamic system of mutualism whereby each trader is dependent on their counterpart to provide items essential for continued success in an area of limited means. The direct and intimate nature of the Chilote barter process nurtures social relations while maintaining the ecological vitality from which it was derived. The persistence of this barter system provides a degree of economic self-sufficiency and regional sovereignty for communities of an otherwise economically underdeveloped corner of the world. The Chilote barter network provides a living link to historic forms of sustainable commerce and resource exchanges that have existed for hundreds and thousands of years. Perhaps even more timely is that it also illustrates the viability of alternative local and regional resource economies that are less vulnerable to global market forces while meeting essential needs for poor households.

Conclusion

This research found that the Chilote barter is still actively practiced. However, the extent to which it is still affectionately referenced and mythologized is much larger than the area and population that actually still engages in the practice. The spatial area of extent is currently limited to trade between the nearest mainland and island communities that produce wood or potatoes. In this case, a range of 60–100 km between trading partners was found to be the geographical extent of the trading network. This was due primarily to the transportation capability and safety of the wood-traders' boats for navigating the challenging inner seas of northern Patagonia. A population of approximately 30 wood traders and 50 potato growers are those that were identified during the course of this study.

The distinct ecological characteristics of the two areas that make up the trade network are what largely drive the Chilote barter system. The mainland forest ecosystem has low horticultural productivity levels but high productivity in the form of woody biomass due to the intact native forest, high precipitation levels, and poor soil quality. The Comau Peninsula's inability to produce a stable food source is the most important ecological factor driving the participation of the wood-traders in the Chilote barter. All of the wood producers interviewed explained they had a small potato patch on their land, but that it was never productive enough to exclusively provide for the family and was often prone to failure due to soil saturation, molds and fungus outbreaks, and poor nutrient conditions. However, the dense coastal temperate rainforests of the peninsula provide ample material for non-horticultural activities such as fishing and trading of wood products.

Meanwhile, the island communities practice a well-developed horticultural-pastoral system that provides a limited surplus of staple foods such as potatoes and

Image 5.4 Sheep, fence, and potato field on Meulín Island. (Photo credit: Richard Vercoe)

sheep. The historic and continued clearing of large woody biomass such as trees makes the islanders dependent on mainland sources for essential land-management resources such as fence posts. The resulting dynamic is a nonmonetary, direct-exchange of regional resources through ongoing trade relationships reinforced through specific cultural practices and behaviors such as the maté tea session (Image 5.4).

The Chilote barter facilitates both an exchange of necessary natural resources and a system of social continuity within and between remote communities. Both parties honor preexisting exchange rates but recognize that the final exchange rate may vary according to how the participants evaluate the individual products to be traded and the personal needs of each participant at the time. The final satisfaction of the transaction is achieved not only by obtaining the desired material resource but also by fulfilling the related social needs as well. The 10 posts for a 50 kg sack of potatoes exchange rate is identified by all of the traders as a long-standing base-rate upon which every negotiation begins. This could be called an equilibrium exchange rate, a balance between what the natural system consistently provides and that of the human resources and labor involved in cultivating and transporting the goods. This exchange rate is reflective of a natural carrying capacity of the socioecological system for each of the traded products.

However, the traditional exchange rate is not necessarily the final exchange rate. The intrinsic social dynamics of barter must also be factored into each exchange. A barter must be viewed in light of its social context as well as its material practice (Humphery and Hugh-Jones 1992). Due to the personal nature of bartering, each participant's reputation is at stake for future trades. Therefore, a *socially* evaluated exchange rate is also present in any Chilote barter process. Thus, the exchange rate considers the social value associated with the products in addition to their labor and use values. Examples of the social value include familial ties or religious/ceremonial connections between traders who attend major annual pilgrimages and holy days together.

Moreover, the "days exchanged" labor invested by members of each community to each family's potato harvest or boatload of wood creates a social accountability that both traders are intimately aware of. For example, during an observed barter negotiation, Don Samuel explained to Don Pato that he could not offer 12 posts for a sack of potatoes because he already traded with neighbors Don Sixto and Don Oscar for 10 posts per sack and it would jeopardize his future opportunities with the others if he accepted something different during the same visit. Such social considerations are a factor in the evaluation of the quality of the products being traded. In other words, the ultimate valuation of the wood or potatoes being exchanged is based not only on their natural qualities but also on their social qualities at the time of trade. The Chilote barter demonstrates an intrinsic awareness by the participants of not only an ecological equilibrium of the products traded but also an attunement of social relations that are continually cultivated and maintained by the exchange process.

References

Cellarius, B. 2000. You can buy almost anything with potatoes: An examination of barter during economic crisis in Bulgaria. *Ethnology* 39: 73–92.

FAO. 2011. *Beijing declaration: A ten-point charter to promote the dynamic conservation of globally important agricultural systems (GIAHS)*. Crab Island, China: United Nations.

Humphery, C., and S. Hugh-Jones. 1992. *Barter, exchange,and value: An anthropological approach*. New York: Cambridge University Press.

Nahuelhual, L., P. Donoso, A. Lara, D. Nuñez, C. Oyarzún, and E. Niera. 2007. Valuing ecosystem services of Chilean temperate rainforests. *Environment, Development and Sustainability* 9: 481–499.

Sauer, C. 1969. *Agricultural origins and dispersals*. Cambridge, MA: M.I.T. Press.

Yao, S. 2010. True origins of widely used potato germplasm revealed. *Agricultural Research* 58: 22–22.

Chapter 6
Seeing the Forest for the Trees: The Firewood Trade in Southern Chile

Eric H. Thomas

Introduction

During the summer, the Patagonian town of Puerto Aysén boasts spectacular views of the mountains that surround it and of the plain that runs west toward the great fjord to the Pacific. Puerto Aysén is a frontier town and regional transportation hub linking the seafood plants, free trade zone, and ferry terminal at Puerto Chacabuco with the regional capital at Coyhaique. Its orange suspension bridge, completed in the middle of the last century, remains one of the most iconic in Chile. In the winter, however, the town feels different. Low clouds settle over Puerto Aysén for weeks at a time, obscuring the mountains and bringing a near constant drizzle. Water permeates the shingled wooden houses and rattles against their corrugated roofs. The air is thick with the smell of wood smoke and flatbed trucks with stacks of firewood harvested from the hills outside of town trundle over the bridge.

If you walk northwest from town, you will soon come upon a different bridge. This too is a suspension bridge, though it is much smaller than the one in town, and it connects the settlement at Los Palos—a collection of farmhouses and small holdings—to the larger community at Puerto Aysén. Next to the bridge there is an unassuming monument. Dedicated in 2012, it is a wooden sculpture of a *leñero*, a firewood trader, rowing his boat down the Rio Los Palos, on his way to Puerto Aysén. The engraving below the sculpture reads, in part:

> The first inhabitants of the community were colonists who arrived principally from the island of Chiloé. The economic activity in which they participated was the selling of firewood, [an activity] in which the whole family participated and in this way cleared their fields to make pastures and raise cattle. When the road was the river, the wood was transported in boats to the nascent city of Puerto Aysén, where it was sold to the inhabitants and to the ships of the era, which used it as fuel to produce steam. With their earnings [the

E. H. Thomas (✉)
Department of Anthropology, University of North Carolina, Chapel Hill, NC, USA
e-mail: ehthomas@live.unc.edu

A. Daughters, A. Pitchon (eds.), *Chiloé*, Ethnobiology,
https://doi.org/10.1007/978-3-319-91983-6_6

Image 6.1 The monument at Los Palos depicting a *leñero*. (Photo credit: Eric Harrison Thomas)

colonists] bought necessary provisions and returned to their fields taking advantage of the tidal differences to come down the river and to re-ascend it on their return.[1]

Since these first inhabitants settled along the fjord in the early twentieth century, "those who arrived first," as the monument states, traded firewood. Despite the rapid development of the region beginning in the 1970s, this ongoing trade in wood harvested from the region's many hardwood forests remains vital, sustaining residents during the long, wet, Patagonian winters. Recent state interventions in the industry, ranging from efforts to regulate the quality of firewood to initiatives to subsidize its cost, attest to the continued importance of firewood in the day-to-day life and governance of the Aysén region. New concerns about air quality in cities like Coyhaique, however, are beginning to affect this hundred-year-old industry, perhaps changing how residents think about firewood—and by extension, their longstanding traditions (Image 6.1).

Research for this article was conducted over a 4-month period during the austral winters of 2015 and 2016. The research combined traditional ethnography,

[1] Translation by the author.

observational research, secondary data analysis, and survey research. The methodology for this project included semi-structured interviews with 12 *leñeros* (firewood traders) as well as several government officials living in the region. Additionally, a survey of 30 households was conducted in July 2015 to assess household-level responses to the Chilean state's introduction of a subsidy in 2012 and subsequent market fluctuations.[2]

The *Leñeros*

If you ask anyone living in the Aysén region what sets it apart from the rest of the country, the response you're most likely to get is "*la leña y el mate*" (firewood and yerba mate tea). The latter attests to the presence of the Patagones who, usually with herds of livestock, settled the interior of the region by coming over the mountains from Argentina. The former, however, unites these settlers with the Chilotes who came from the islands of Chiloé by sea. Today, these two groups along with the thousands of new arrivals drawn to the region by the booming aquaculture industry all depend on firewood. The modern-day *leñeros*, descendants of the stoic figure depicted in the monument at Los Palos, are the backbone of this essential industry. Their environmental knowledge (Stone 2016) enables them to operate successfully, even during the depths of the Patagonian winter.

Patricio,[3] one such *leñero* who sells in Coyhaique, described the qualities that defined the best firewood and therefore commanded the highest prices. The best firewood is always the *más duro* (hardest) and *más liviana* (lightest) because that indicates that it has been properly cured. But for *leñeros* and their customers, there is no scientific way to measure how dry the wood is. Instead, according to Patricio: "one knows just by looking, because the wet wood is a different color. It's darker. When the wood dries it shrinks, so you recognize this too. Obviously also if you touch it you can tell."

Though he trades in firewood, Patricio insists that he can identify the quality of firewood because he is *from* Aysén. In his words, "it's something cultural." Though they were not *leñeros,* several generations of men in his family cut their own firewood in the valley around Rio Simpson, near Coyhaique. Patricio learned what qualities made for good firewood when he accompanied them as a boy.

For Patricio, as with most *leñeros* operating throughout the region today, work in the firewood trade represents only one part of a multi-occupational career. Though he has transported and sold firewood full-time for the past three years, prior to this

[2] Research was funded through a Tinker Foundation Field Research Grant provided by Institute for the Study of the Americas, a Thomas F. Ferdinand Summer Research Fellowship from the Graduate School, and a Harriet J. Kupferer Graduate Investigators Grant from the Department of Anthropology at the University of North Carolina, Chapel Hill.

[3] A pseudonym. With the exception of government officials who granted explicit permission, all names that appear in this chapter have been changed to protect the anonymity of participants.

he worked as a low-level administrator in the municipal government in Coyhaique, looking after the city's parks and monuments. During this time, he sold firewood part time to people he knew in and around the city. Patricio sells logs that have already been cut down and split to fit into a woodstove, as do many part-time traders. These *leñeros* act more like middlemen, buying large quantities of wood from other *leñeros*—who transport and sell logs in meter-long segments on flatbed trucks from more remote parts of the region—and then cutting it down before reselling it, effectively adding value to their product.

Other part-time workers are able to earn extra cash by acting as "ride-alongs" without making a significant investment in their own equipment. They help load and offload firewood from the flatbed trucks of friends and relatives. Still more residents earn extra cash by cutting down the meter-long segments brought back by larger operators. These men are a near constant sight in Puerto Aysén, where they travel with chainsaws on the backs of their bicycles, in many cases following the larger flatbeds and offering their services to local buyers as soon as meter-long logs have been offloaded.

Most *leñeros* today do not cut down the trees that produce firewood themselves, but collect them from designated pick up sites with road access. Families in the countryside fell the trees and cut them into meter-long segments, coordinating with the *leñeros* by cell phone to let them know when and where a load is waiting. These families, who are typically harvesting by clearing land on their own parcels—much as their forebears did—are the ones who identify the *lenga, ñire,* and *roble* beech trees (Nothofagus pumilio, Nothofagus Antarctica, and Nothofagus obliqua respectively) that are sufficiently old to produce the kind of thick, hard logs that burn hot and "clean" (Image 6.2).

On the nearby Islas Huichas archipelago, home to fewer than 2,500 year-round residents, the process of procuring firewood is no less essential but considerably more complicated. Since the first Chilotes arrived and established Puerto Aguirre—the largest port on the islands—in 1942, nearly all of the *lenga*, *ñire,* and *roble* on the inhabited islands have been cut down. This means that suitable hardwood must be harvested from nearby uninhabited islands and brought back to the archipelago by sea. As with Patricio and other mainland *leñeros*, for those trading firewood on the islands it is one of several jobs, facilitated by owning and operating a boat (or working as a ride-along). This means that many fishermen bring their catch back in the late morning or early afternoon, only to depart again to bring back a load of firewood late in the day.[4] During the winter, when the weather makes fishing more difficult, this is a valuable and considerably more reliable way for fishermen to pay their bills—though there is considerable danger in transporting meter-length logs over rough seas.

[4] Since Chile instituted a quota system within its fisheries, many of the islands fishermen have given up subsistence altogether, and sell their quotas directly to the industrial fishing firms operating in the region. This has given them greater flexibility in terms of the hours in which they are able to transport firewood from nearby islands.

Image 6.2 Firewood brought by sea to the community of Puerto Aguirre in the Islas Huichas. (Photo credit: Eric Harrison Thomas)

Acquiring Firewood Knowledge

The *leñeros*, whether they are operating on the mainland or the islands, are not the only ones for whom a kind of firewood knowledge is essential. Chileans living in Aysén—whether they were born and raised in the region or, like those overseeing salmon production, came more recently in search of economic opportunities on the frontier—must acquire a host of practical skills in order to heat their homes. Firewood knowledge is therefore not monolithic. For those who earn money trading in firewood, thereby converting it from natural resource into a commodity, this knowledge is expansive, and includes knowledge of the forest, of regional roads, of shifts in the weather, and of how many logs can be safely transported by sea. For those who buy it, though this knowledge is narrower and more likely confined to the domestic sphere, it is no less important.

For everyone in Aysén, firewood knowledge informs day-to-day practices, shapes identity, and creates bonds of solidarity. What it takes to build a fire, to keep it burning—and burning safely—is essential knowledge that can only be acquired by practice. As such, the skills associated with firewood represent a kind of *métis*, or knowledge that can only be gained by engaging in the activity itself. Firewood knowledge in Aysén therefore consists of "a wide array of practical skills and acquired intelligence in responding to *a constantly changing natural and human environment*" (Scott 1998:313, emphasis added).

Firewood knowledge requires not only an understanding of the materials and practices required to heat one's home effectively and safely (knowledge of the natural) but also, and perhaps more importantly, a thorough understanding of the social relations entangled in the procurement and use of firewood (knowledge of the social). Firewood is produced through human actions—certainly firewood is different than deadwood or living trees—but it also *produces* certain types of people and the social relations among them. Firewood knowledge and practices mark who belongs and who does not.

For those like Patricio, who have grown up in the region, firewood knowledge develops through a lifetime of experience. But the region's population has grown dramatically since the early 1980s, fueled largely by the booming aquaculture industry (Schurman 2004). Today, Chile produces nearly a third of the world's farmed salmon (FAO 2008) and despite the recent ISA (Infectious Salmon Anemia) outbreak that hit Aysén from 2007 to 2010 (Gerhart 2017; Barton and Fløysand 2010) and the contamination of the Chiloé fisheries in 2016 (Fernando González 2016; Daughters 2016; Marín 2016), the cultivation centers and processing plants that depend on aquaculture are some of the largest employers in the region. Nearly 6,000 people living in Aysén are directly employed in the aquaculture sector and another 3,000 gain indirect earnings from the industry (Gobierno Regional de Aysén 2009).

Many of the new arrivals that have been drawn to the region by the promise of work come from northern regions where the use of firewood is rare, if it occurs at all. While those who have grown up in Temuco, Puerto Montt, or on the Chiloé archipelago may arrive with experience, those from farther north may never have operated a woodstove. How do they acquire the kind of firewood knowledge that keeps them safe and warm during the winter months?

It is difficult to overstate the amount of manual labor that goes into heating a home with firewood. As noted in the previous section, firewood can be delivered in meter-length segments or in shorter pieces that can fit into a woodstove without being cut down. If residents take delivery in meters, they have to either use a chainsaw to cut it into three or four smaller segments or hire someone to do so. Once this has been accomplished, they have to split the pieces of appropriate length. This is done with an axe and it is exhausting and dangerous work.

Leaving aside *building* the fire, which is less labor intensive but which—as any novice soon realizes—presents a challenge (Henne 2010), anyone heating their home with a woodstove must regularly check on the fire at intervals that depend on the stove and the quality and size of the firewood it consumes. Fire, in both its production and its maintenance, is indeed "extraordinarily interactive" (Pyne 2001:46).

Florencia, an architect who had to learn firewood skills when she moved to the region from the northern city of La Serena, described the daily conversations she had with her husband: "Did you feed the fire? How is the fire today? It's like having another person there!"

Florencia made her complaint at a dinner party with guests who had grown up in the region *and* with relative newcomers. While Florencia's frustration with the need to tend to her woodstove may have marked her as an outsider (had she grown up in Aysén none of this would have struck her as odd), the affirming nods she received from around the table suggested solidarity based on shared labor and shared experience. As long as the discussion was about chimneys, the price and quality of firewood, and the work involved in keeping the fire going, everyone could participate equally. Everyone in attendance had labored to split logs in inclement weather and felt frustration when an improperly cured log dampened the fire just as they were trying to dry clothes or toast *marraqueta* (a type of Chilean bread), along with dozens of other shared experiences that could only come from relying on firewood.

The assembled guests had also likely shared the experience of an *incendio* (house fire), if not in their home or on their street then certainly in their neighborhood. These fires are a regular occurrence in Aysén. In a region where nearly every household relies on a woodstove for 9 or 10 months out of the year, this is perhaps not surprising, yet we should remember that the only thing that keeps these disasters from being even *more* common is firewood knowledge.

These kinds of disasters are never far from the surface. When I was in the region in the winter of 2015, four homes in Puerto Aysén burned, three of them on the block where I lived. Of 30 families surveyed for this project, two-thirds reported that they either knew or were related to someone who had lost their home to fire, and two reported that they had lost *their own* homes to fire. Data provided to me by the superintendent of the local volunteer fire department tracked 180 household fires in Puerto Aysén (a town with a population of roughly 15,000), at least 33 of which began because of creosote buildup in the chimney *in 2014 alone*. According to the superintendent, more than 70% of the fires that occur in Puerto Aysén are caused by wood burning stoves, either because residents fail to clean their chimneys regularly or because they do not dispose of the ashes generated by their woodstoves safely, causing unnoticed hot coals to ignite other materials.[5]

This highlights another fact of life on the frontier: the failure or inability to acquire and mobilize firewood knowledge can be deadly. Though Aysén is not unique for the wood-frame construction of most houses (this is common farther north in cities like Temuco and Puerto Montt as well as in rural parts of the Región de los Lagos) this building technique, the proximity of houses to one another in villages and towns and the reliance on wood burning stoves do create a higher risk of fire than is found anywhere else in the country (Image 6.3).

If demonstrating local firewood knowledge creates bonds of solidarity, the failure to do so—or even the appearance of failure—creates friction. The risk of fire is a major factor in disputes between neighbors. Several survey respondents told me how concerned they were about their neighbors, either because they were irresponsible or, in one case, because the neighbor in question was allegedly an alcoholic and therefore couldn't be trusted to take care of her stove safely. In Puerto Aysén, where fires can and do occasionally burn whole city blocks, these concerns are hardly unfounded, even if they are presented with the kind of middle-class sensibility typically associated with busybodies who worry about how often their neighbors cut their lawns or weed their gardens. Yet this demonstrates how the sense of vulnerability is *shared*. The informal "policing" of neighbors by neighbors represents a kind of collective action for the greater good of the community.

[5] Fires are so common that the cost of insuring your home against this kind of disaster is prohibitive for all but the wealthiest residents.

Image 6.3 Interior of a home in Puerto Aysén destroyed by fire in 2015. (Photo credit: Eric Harrison Thomas)

A Chaotic State Intervention

The firewood industry in Aysén, which had operated in much the same way for generations, was completely upended in 2012, when the Chilean government introduced a subsidy to be administered by the *Fondo de Solidaridad e Inversión Social* (Fund for Solidarity and Social Investment, more commonly referred to by its acronym, FOSIS). State officials announced the subsidy in the wake of protests against the proposed HidroAysén project (Varas et al. 2013; Zibechi 2012), which would have installed five dams on the Aysén region's Rio Pasqua and Rio Baker. Residents were outraged, not least because nearly all of the 3,500 MW/yr generated by the project would go to Santiago, 2,400 km to the north (Vince 2010) while electricity in Aysén would remain the most expensive in the country. Thus, the distribution of firewood by FOSIS to "vulnerable" families throughout the region seemed like a good way to placate residents with grievances over the high cost of living on the frontier.

In Aysén, the regional director of FOSIS is a short, intensely energetic woman named Claudia. She describes her staff as always looking for ways to be "closer to the people" in order to work "on the dynamic problems experienced by families." When her staff was charged with implementing the firewood "bonus" for residents of Aysén in 2012, everyone in her office knew that this would be challenging. First, this would be a new subsidy, unique to Aysén, and thus there was no national precedent to draw on for guidance. Second, the subsidy would target families deemed "vulnerable" via a complex system of "points," but not all of these families were in the FOSIS database, and many of those who were lived in places so remote as to be inaccessible for the large trucks that transported firewood during the winter months. Finally, FOSIS would work with the regional government to buy up and redistribute firewood, but they had no one on staff with experience cutting or transporting firewood.

The result, according to Claudia, was "chaos." Initial attempts to buy and redistribute firewood to 16,000 qualifying families throughout Aysén had disastrous consequences. With the government buying up large quantities of firewood, prices skyrocketed, jumping from $13,000 Chilean pesos per cubic meter (roughly US$24 in 2012) to $30–33,000 pesos per cubic meter (between US$48 and US$52 today). The results were predictable: middle-class families who didn't qualify for the subsidy suddenly found themselves with *higher* heating costs, and many residents who had previously worked in other industries rushed to enter the firewood market, either cutting wood or transporting it.

Despite the obvious setbacks—and massive dissatisfaction on the part of the local populace—FOSIS pressed on with its mandate to provide the subsidy, and in so doing displayed remarkable flexibility. Going into the subsidy's second year, the agency switched from attempting to buy and redistribute firewood to providing *cash* bonuses of $100,000 pesos (roughly US$167) with which residents could buy their own firewood from traditional suppliers. They also continued to seek out and enroll new qualified families,[6] expanding the initial enrollment from 16,000 to 23,690 families by 2014. This new system, providing cash rather than firewood itself, has led to new challenges and to strange sights: the day after our interview, Claudia traveled with the regional manager of Banco Estado, two cashiers from the bank, three *carabineros* (police officers), and an armored truck to the island community at Melinka (population 1,400) by ferry in order to distribute subsidies there. The commotion caused by this veritable "army" in a small island community, where residents lined up to receive their payments, cannot be overstated.

But for all the "chaos," or perhaps because of it, the introduction of the firewood subsidy in 2012 provided residents, including those who did *not* qualify for the subsidy—whether they were Ayseninos or recent arrivals—with opportunities to tactically shift their practices in order to profit from[7] the state's intervention.

[6] They even rolled out a new section of their website where residents can enter their RUT (Chilean national identity number) and see if they qualify, though Claudia freely admitted that this was little help when it came to providing services to families who lacked internet access either because they could not afford it or because they were older and had never learned to use the internet—or in some cases what it is.

[7] The word that most used was "*aprovechar*" a verb that means "take advantage of," "use," "leverage," or "make the most of" depending on context.

"A Cooperative Culture"

In response to the introduction of the subsidy, or more accurately, to a state agency upending the firewood market, many residents who had previously worked only part-time cutting or transporting firewood embraced new strategies and scaled up their operations. FOSIS's efforts to buy and redistribute unprecedented amounts of firewood had expanded state influence but also created a space in which local *leñeros* could operate tactically, taking advantage of oversights, errors, or loopholes in official policy in order to benefit from new conditions. Since this was an industry with which state agents had no previous experience, oversights abounded.

In Coyhaique and Puerto Aysén, groups of firewood transporters had already come together to form what they termed "cooperatives." This was initially in response to the state's limited efforts to regulate the *quality* of firewood through a network of NGOs, government agencies, and the Corporación Nacional Forestal—Conaf (Conway 2010; Henne 2010). Conaf is a semi-autonomous state agency tasked with regulating the timber and firewood industries *and* operating Chile's national parks and reserves. In Aysén, Conaf had already introduced new measures designed to ensure that firewood being sold was dry—since damp wood produces more smoke—but these were hard to enforce and had little effect. In the wake of the subsidy, however, the region's firewood cooperatives took off. One such cooperative was *Coleña Aysén*, whose co-founder, a middle-aged man named Gilberto, still wears the black *Coleña* baseball cap years after the cooperative's heyday and subsequent decline. Gilberto was *Coleña*'s president during the organization's most successful period from 2012 to 2013, when the sudden jump in prices and the need for ride-along laborers, who do not own trucks themselves, encouraged men who had previously worked only part-time to join the cooperative. When the subsidy was announced and FOSIS began to contract with the cooperatives in order to buy in larger volume, Gilberto and the men of *Coleña Aysén* sold more than 2,000 cubic meters of firewood *in the first ten days* of their contract with FOSIS.

Today, FOSIS's strategic shift from the redistribution of firewood to the dispersal of cash bonuses, in order to "sustain and optimize" their process (Li 2007:19), has diminished the need for and profitability of these cooperatives. As we stood beside his two flatbeds on the *ribera sur* (south bank) of Puerto Aysén, Gilberto estimated that only 50% of those transporting and selling firewood in Puerto Aysén belonged to the cooperative in 2015. During our conversation he told me with ambivalence, "in Chile we do not have a cooperative culture," yet the short-term success of the cooperatives in mobilizing to take advantage of the FOSIS subsidy in 2012 and the "chaos" it brought to the firewood market are striking. Rather than chafing under government restrictions or policies that affected how they could sell firewood, many local actors within the industry responded quickly and in novel ways to meet new challenges and to improve their economic well-being. This response was surely connected to the need for improvisation that has historically been an essential part of living on the frontier (Bengoa 2003; Gomez 2002), as well as to the informal networks that have bound the community together since before state attempts to "modernize" the frontier. Furthermore, while the *country* may not have a cooperative culture, responses like those of Coleña Aysén demonstrate that the *region* does.

Though he is no longer president of the cooperative, Gilberto has taken advantage of the subsidy to improve his standing within Puerto Aysén and to enhance his own economic position. While previously he traveled with only one truck, he used the money he made during the early days of the subsidy to buy a second flatbed and he frequently borrows a third from a friend in order to take a "convoy" to the rural sites where wood is cut and stacked. With three men in each truck, he is able to bring back between 50 and 100 cubic meters per trip, and then is able to store this in an *almacén* (storehouse) outside of town and sell it at his own pace—and to take advantage of seasonal price fluctuations that affect those who do not have enough space to store sufficient firewood to get through the winter. He is not alone in this approach. Since 2013, the most successful *leñeros* in Puerto Aysén have nearly all expanded or built new storage facilities. This strategy—making fewer trips with more workers to bring back larger quantities—reduces Gilberto's risk during the winter months when the weather and local roads are unpredictable and keeps the journey, which smaller operators must make every 10–15 days, from becoming what he calls "an adventure."

A Strategic Return to Older Practices

Beyond encouraging new collective economic formations like the cooperatives, the introduction of the FOSIS subsidy also encouraged a return to *older* practices of living and operating on the frontier—what I will term "strategic traditionalism." Some residents embraced "salvage rhythms." This "unregularized coordination" that relies on natural temporalities (Tsing 2015:132–133) describes how some Ayseninos opportunistically gather firewood from the sparsely populated countryside during the summer and fall. Gualterio, who has lived in the region for a decade and who studied forestry (and who has a number of strong opinions on the state's regulation of the timber industry and maintenance of reserves and parks), used his local knowledge to avoid the "chaos" of 2012. When I asked him about the subsidy, he informed me matter-of-factly that he thought the state was responsible for the current price in firewood and lamented how the massive price increase caused by the FOSIS intervention meant that anyone able was selling *any* firewood they could get their hands on, regardless of its quality. This, he said, effectively undermined any efforts to ensure that only cured wood was being sold and burned, since no one wanted to wait to cure the wood so that it would burn cleanly and safely.

Gualterio showed me the bed of his *camioneta* (pickup truck), which was full of gnarled wood that he had collected. Though it wasn't cut into the flat, straight, meter long segments found on the back of the *Coleña* trucks, it was indeed very dry. "The forests around here are full of this stuff," he said, but Conaf's policy regarding its collection remains something of a "gray area." Official policies have restricted access to the region's reserves and parks, but this is, in his opinion, only because it is easier for them to monitor (and profit from) the cutting of new stands of trees than the recovery of deadwood from the forest floor. Gualterio cited the risk of wildfires as one of the reasons he thought it was irresponsible for Conaf not to allow individuals to collect what had fallen naturally to the forest floor.[8] When I asked him if he had gone into the reserves and parks to collect what he had in the back of his truck he laughed and shook his head, telling me cryptically that he goes only to "places he knows." For Gualterio these are not places of capitalist accumulation, since he is collecting for himself rather than to sell. For many *leñeros*, however, spaces like these are simultaneously inside and outside of contemporary capitalism (Tsing 2015:63), just as they have been since the first colonists arrived in Aysén.

In cases where residents have neither the time nor the resources to salvage deadwood from the countryside, they have embraced other traditional practices to offset the soaring cost of heating their homes. The most common of these is multigenerational living, with three generations under the same roof. Having multiple generations in the same small house has always been more common in Aysén than in less remote parts of the country, but moving extended family into a shared house has become strategic since the introduction of the FOSIS subsidy. Because most elderly people collecting a pension qualify for the subsidy, this has become a means for adult children (who are typically responsible for the care of elderly parents anyway) to defray heating costs for themselves and their own children even if their income level is not low enough for them to receive the subsidy.

This strategic return to practices that had become less common on the modernizing frontier is not intended to subvert the new state-ordered system but merely to adapt to it by returning to what once was. Whatever strategies they have embraced, whether novel or traditional, both Ayseninos and recent arrivals suffered as a result of the subsidy and both sought to mitigate their susceptibility to the price increases that accompanied it. Even those who were not originally from Aysén embraced strategies—like Gualterio, who salvaged deadwood—that have been practiced in the region since before its formal incorporation into the Chilean state. Furthermore, the acquisition and deployment of firewood knowledge has long shaped identity in Aysén, but the consequences of a policy implemented by officials who clearly *lacked* that knowledge drew proficient new arrivals and Ayseninos closer together.

[8] This risk is very real. Portions of Torres del Paine National Park, just south of Aysén in the Magallanes Region, burned because of fires by backpackers in 1985, 2005, and 2011. In January 2017, the worst wildfires in Chilean history burned some 900,000 acres in the middle of the country. This was the result of an explosive combination: long-term drought, unprecedented summer temperatures, and changing land use patterns.

New Battles over Contamination

Though the industry has stabilized in the five years since the subsidy was introduced, today the firewood trade once again finds itself the subject of heated debate. In 2016, the World Health Organization published an updated version of their Ambient Air Pollution Database, which tracks the airborne particulate levels over cities around the world. Perhaps not surprisingly, many of the cities with the worst air quality can be found in the rapidly industrializing states of southern and southeast Asia. What *is* surprising, however, at least for those living elsewhere, is that Coyhaique—the sleepy regional capital of Aysén—tops the list of South American cities included in the database with an average annual PM 2.5 level of 64 $\mu g/m^3$ (WHO 2016).[9] The air quality during the winter months has become such a concern that the Chilean Ministry of the Environment and Ministry of Health now jointly run a website that monitors daily air quality, warns residents against "physical activities of medium and high intensity," and prohibits the operation of industrial boilers between certain hours (Gobierno de Chile 2016).

Those who live in Coyhaique during the winter months, when the cold temperatures and surrounding mountains trap smoke over the city, were not surprised by the WHO findings. Here, the smell of smoke permeates the air, saturates clothing, and eventually causes the kind of scratchy throat and dry eyes normally associated with allergies. Asthma rates have risen steadily as the region's population—and consequently the number of wood-burning stoves—has grown.

The WHO findings, along with ongoing conversations about the air quality in Santiago,[10] have sparked intense discussion among those living on the southern frontier. While firewood remains the only home heating option for most, residents are aware of the dangers and side effects of its continued use.

And yet, the situation seems unlikely to change anytime soon. In 2015, José Urrutia, head of Forest Development at Conaf called firewood "perhaps the most democratic product there is." The firewood industry, he explained, provides employment for at least seven hundred families across the region. In contrast, the electric and gas industries in Chile operate more or less as monopolies, with only 2–3 companies reaping the profits from sales in the country's 15 regions.[11] Were a viable fossil fuel alternative to firewood to emerge, even *more* long-time residents would find themselves out of work.[12] They, like many of their neighbors, would be forced

[9] According to the WHO annual mean levels of PM 2.5 above 35$\mu g/m^3$ are associated with significantly higher risk of mortality.

[10] Another city surrounded by a ring of mountains where the contamination is bad enough that certain districts, though not the affluent *comunas* of Los Condes and Vitacura, have outlawed the burning of firewood altogether.

[11] Quoted in *El Divisadero*, p. 24, Año XXII, No. 6286 July 1, 2015.

[12] Indeed, the general lack of fossil fuels within Chile's borders (particularly compared to neighboring Argentina) has long been considered a threat to national security, and the mining of relatively limited coal deposits around Lota has played on outsize role in Chilean politics (see Pavilack 2011).

to work full-time in the region's seafood plants, where the hours are long, the work is monotonous, and the wages are unreliable (Gerhart 2017; Schurman 2004). Work in the firewood trade, particularly after the drastic price increase that accompanied the introduction of the FOSIS subsidy in 2012, remains a better option for those seeking a greater return on their investment and greater independence.

Seeing the Forest for the Trees

The firewood trade has been the backbone of Aysén's economy since the first Patagones came through the mountain passes that connect it to Argentina and the first Chilotes picked their way down its fjords. Even now, as Chilean politicians tout the need for a diversified economy built around export products for foreign markets (Foxley 2004; Valdez 1995), this intensely local industry provides valuable income for many and an invaluable resource for nearly everyone, whether they are Ayseninos born and raised in the region or new arrivals. Since the region is home to nearly 35% of Chile's native hardwood forests but less than 1% of the country's population, the trade does not appear to be in any danger, though *leñeros* must go farther and farther from the region's cities and towns to collect their product. For now, the very conditions that make firewood a necessity—the region's rugged coastline and steep Andean slopes—also limit the expansion of pine plantations and other export-oriented industries and preserve native forests for local use.

Recent concerns with air quality, however, combined with the success of pine plantations in the X Region (Henne and Gabrielson 2012; Klubock 2004) and the ascendance of aquaculture (Gerhart 2017; Barton and Fløysand 2010) suggest that this resource that defines life on the southern frontier may not endure indefinitely. Today, when you ask Ayseninos what sets their region apart they will likely answer, as they have for nearly a century, "*la leña y el mate*." What they will say in another 50 years, however, is less certain.

References

Barton, Jonathan R., and Arnt Fløysand. 2010. The Political Ecology of Chilean Salmon Aquaculture, 1982-2010: A Trajectory from Economic Development to Global Sustainability. *Global Environmental Change* 20: 739–752.

Bengoa, José. 2003. 25 años de estudios rurales. *Sociologias, Porto Alegre* 5 (10): 36–98.

Conway, Frederick J. 2010. Firewood Certification in Chile: Equity in an Innovative Form of Alternative Trade. *Human Organization* 72 (1): 55–64.

Daughters, Anton. 2016. Fish Kills and Protests on the Islands of Chiloé. *Anthropology News* 57 (5): 61–66.

FAO. 2008. *World Review of Fisheries and Aquaculture*. Rome: United Nations Food and Agriculture Organization http://www.fao.org/docrep/011/i0250e/i0250e00.htm.

Fernando González, Pablo. 2016. Chiloé a pasos de convertirse en Chernobyl. *Le Monde Diplomatique, Edición Chilena*. February 20. http://www.lemondediplomatique.cl/CHILOE-A-PASOS-DE-CONVERTIRSE-EN.html. Accessed 25 Nov 2016.

Foxley, Alejandro. 2004. *Successes and Failures in Poverty Eradication: Chile*. Washington, DC: The International Bank for Reconstruction and Development/The World Bank.

Gerhart, Andrew. 2017. Petri dishes of an archipelago: the ecological rubble of the Chilean salmon farming industry. *Journal of Political Ecology* 24: 726–742.

Gobierno Regiónal de Aysén. 2009. *Estrategia Regional de Desarrollo de Aysén*. Chile: Coyhaique.

Gobierno de Chile. 2016. Seremi del Medio Ambiente. *Todos Descontaminamos Coyhaique*. http://airecoyhaique.cl. Accessed: March 14, 2016.

Gómez, Sergio. 2002. *La 'Nueva ruralidad': ¿Qué tan nueva?* Santiago: LOM Ediciones.

Henne, Adam. 2010. Green lungs: good firewood, healthy air, and embodied forest politics. *Environment and Planning A* 42: 2078–2092.

Henne, Adam, and Teena Gabrielson. 2012. Chile is Timber Country: Citizenship, Justice, and Scale in the Chilean Native Forest Market Campaign. In *Environment and Citizenship in Latin America: Natures Subjects and Struggles*, ed. Alex Latta and Hannah Wittman, 149–166. New York: Berghahn Books.

Klubock, Thomas Miller. 2004. Land, Labor, and Environmental Change in the Forestry Sector in Chile, 1973-1998. In *Victims of the Chilean Miracle: Workers and Neoliberalism in the Pinochet Era, 1973-2002*, ed. Peter Winn, 337–388. Durham, NC: Duke University Press.

Li, Tania. 2007. *The Will to Improve: Governmentality, Development, and the Practice of Politics*. Durham, NC: Duke University Press.

Marín, Veronica. 2016. Industria salmonera ha perdido US $80 milliones por manifestaciones en Chiloé. *El Mercurio Online*. May 10. http://www.emol.com/noticias/Nacional/2016/05/10/802062/Industria-salmonera-ha-perdido-US80-millones-de-dolares-por-manifestaciones-en-Chiloe.html. Accessed 22 Nov 2016.

Pavilack, Jody. 2011. *Mining for the Nation: The Politics of Chile's Coal Communities from the Popular Front to the Cold War*. University Park, PA: Pennsylvania State University Press.

Pyne, Stephen J. 2001. *Fire: A Brief History*. Seattle, WA: University of Washington Press.

Schurman, Rachel. 2004. Shuckers, Sorters, Headers, and Gutters: Labor in the Fisheries Sector. In *Victims of the Chilean Miracle: Workers and Neoliberalism in the Pinochet Era, 1973-2002*, ed. Peter Winn, 298–336. Durham, NC: Duke University Press.

Scott, James C. 1998. *Seeing Like a State: How Certain Schemes to Improve the Human Condition Have Failed*. New Haven, CT: Yale University Press.

Stone, Glenn Davis. 2016. Towards a General Theory of Agricultural Knowledge Production: Environmental, Social, and Didactic Learning. *Culture, Agriculture, Food, and Environment* 38 (1): 5–17.

Tsing, Anna. 2015. *The Mushroom at the End of the World: On the Possibility of Life in Capitalist Ruins*. Princeton, NJ: Princeton University Press.

Valdez, Juan Gabriel. 1995. *Pinochet's Economists: The Chicago School in Chile*. Cambridge: Cambridge University Press.

Varas, Pablo, Manuel Tironi, Hugh Rudnick, and Nicolás Rodriguez. 2013. Latin America Goes Electic. *IIEE power and energy magazine*. April 17.

Vince, Gaia. 2010. Dams for Patagonia. *Science* 329: 382–385.

World Health Organization (WHO). 2016. *WHO's Urban Ambient Air Pollution Database Update*. Geneva: Switzerland http://www.who.int/phe/health_topics/outdoorair/databases/AAP_database_summary_results_2016_v02.pdf. Accessed 10 May 2016.

Zibechi, Raúl. 2012. Latin America: A New Cycle of Social Struggles. *NACLA Report on the Americas* 45 (2): 37–49.

Chapter 7
How Households Are Made: Marriage, Independence, and Productivity on the Island of Apiao, Chiloé

Giovanna Bacchiddu

How do households in the most rural parts of Chiloé interact with the natural environment? This chapter presents case studies of several households on the island of Apiao, Chiloé, emphasizing the everyday duties, household responsibilities, and family roles they learn from a young age. Through detailed ethnographic descriptions of a young couple's phases of courtship and marriage, I show how couples slowly separate themselves from their parents' households to become independent, productive units, and synchronize their lives to the island's agricultural cycle. Individuals are valued for their ability to turn their work into productivity and their household into an independent, autonomous unit.[1]

Apiao is a small rural island of 13 square kms, located on the eastern (inland) side of the Archipelago of Chiloé. The nearest town, Achao, can be reached by motorboat within 2–3 h. The island is home to approximately 700 people, mostly of indigenous origin, subsisting on agriculture, shellfish and seaweed collecting, fishing, and small-animal husbandry. A few islanders work for a salary either in the salmon farms adjacent the island or in its schools and small health clinic. However, these individuals still have an agricultural and farming routine to adhere to in their households.

Unlike other islands in Chiloé, Apiao does not have a town center; its houses are scattered across the landscape, each house surrounded by a substantial portion of land and protected by a series of fences. This spatial arrangement implies that

[1] This chapter is based on long-term fieldwork of more than 15 years. This began in 2001 with approximately 2 years of uninterrupted residence on the island. The research is ongoing; recent trips to the field have been funded by the CIIR – Center for Intercultural and Indigenous Research, which is gratefully acknowledged. My deepest gratitude goes to my Apiao friends who have allowed me into their homes and into their lives.

G. Bacchiddu (✉)
Programa de Antropología, Pontificia Universidad Católica de Chile, Santiago, Chile
e-mail: gbacchiddu@uc.cl

A. Daughters, A. Pitchon (eds.), *Chiloé*, Ethnobiology,
https://doi.org/10.1007/978-3-319-91983-6_7

Image 7.1 A view from Apiao showing cleared and cultivated fields. (Photo credit: Giovanna Bacchiddu)

people do not live too close to each other and that each property line encircles a microcosm, a semi-independent unit of production (Image 7.1).

Life in Apiao is organized around, and strictly tied to, its natural environment. Apiao residents divide their time and occupations and continuously shift between the two main poles of land and sea. The land is where people build their houses, where they cultivate their fields and tend their orchards, where they send their sheep and cows to pasture, and where they walk along the dirt road that bisects the island. The sea—visible from most households—is their main reference point in the landscape, the marine space that separates the island from the other inhabited lands of the archipelago, and from the continent; a resource that regularly delivers supplies of edible seashells and algae, and other natural products to be processed and sold. The sea is where divers and fishermen work and where boats reside. The beach is the landing ground for those who arrive and the point of departure for those who leave; pilgrims attending religious festivals meet at the beach and gather there waiting for the miraculous statue of San Antonio to arrive, in order to offer it a proper welcome (Bacchiddu 2011). The double edges of Chiloé's environment shape and define the occupations, lifestyles, and beliefs of its inhabitants; the environment also directly and indirectly determines the island's social dynamics.

Peter Gow (1995) has pointed out that landscape and human relations are strongly imbricated, describing how kinship ties actively shape not just individuals, but the environment. In the following pages, I focus on kinship ties, and particularly on the union of two adults to form a new family, and thus perpetuate society in Apiao.

To understand how Apiao residents reproduce their society, I illustrate how households are formed, and how couples are made. I follow adolescents in their daily tasks and describe the gendered division of labor, explaining that such divisions point to complementarity and practicality rather than exclusivity. I then describe how one couple I knew decided to form a new family. In this way I explore marriage as an assertion of an individual's autonomy and independence, marked by regular production and reproduction of the island's agricultural cycle, pig farming, and delicate system of alliances.

What do Apiao residents think is the proper way to be a man or a woman? In the first part of the chapter I list and illustrate the activities that males and females accomplish within the household or for the benefit of it. I distinguish between male and female tasks, pointing out that in fact most of the activities are shared.

In and around the Apiao Household

Young people, like everyone else in Apiao, are busy with everyday activities in and around their household. These are considered crucial for the well-being of the house and its inhabitants. Boys are expected to accompany, substitute, or complement their fathers in the most physically demanding domestic tasks, such as cutting down trees, chopping wood, clearing the ground of roots, moving heavy loads, fishing, or diving. Girls tend to stay at home and help their mothers or grandmothers with the routine domestic activities. These include feeding the animals, cooking food for the family and the livestock, making bread, doing the washing, carrying wood or water, and attending to visitors. This is the habitual division of tasks when both men and women are present. If men are not present, women will do their work, and vice versa. In fact, people learn nearly all tasks at a young age, and if necessary, men and women will interchange tasks. Moreover, some tasks are performed communally—men, women, adults, and children together.

Typically, male tasks such as cutting and transporting logs require a lot of physical strength and skill. Once cut down with axes, the logs are transported via land with a yoke of oxen, or by sea, using a boat. The logs are then transformed into poles for fencing, or chopped to use as firewood. Men generally do any building and roofing work. Men are also usually in charge of the various types of fences that demarcate fields and property. Fences are typically made using branches from *arrayán* shrubs that are cut, cleared of leaves, and woven, much like a fabric, into fences. Land must be cleared of trunks and roots before being used as potato or wheat fields.

Apiao residents chop wood every day, irrespective of the weather. They prefer to chop small quantities daily rather than preparing and storing a large amount of wood to last the year. Wood is constantly needed to keep the stove going; this is for both heating and cooking, and its fire is extinguished only before going to sleep at night. Men usually chop wood, but not exclusively; all women know how to do it, and do so if necessary.

Men are generally in charge of the sheep, cows, and horses; they tame young animals, and perform operations on them, such as castration. They also oversee the slaughtering of animals, except chickens, which are typically under the purview of women. Men go uphill to the more distant fields to count and check their cows every morning. They yoke the oxen, to use whenever necessary—usually as a means of transport—carefully packing loads on the wooden sleigh attached to the yoke, or to work in the fields. And finally, men go fishing and diving for seashells and crabs. Adult men perform these tasks together with their children, starting at age 5 or 6. Young boys always follow their fathers and anxiously wait to be considered old and good enough to be involved in a task. Boys start chopping wood when they are 9 or10 years old.

Young women help their mothers with the household duties, starting at a young age. They feed the chickens and make a fire in the *fogón*—a small smokehouse near the main house—to cook the food for the pigs. They collect dry wood for the fire, prepare the food, and look after it. Usually several panfuls must be cooked, as the pigs need food twice daily. Women are in charge of fetching water: they often need to walk a distance to a pit, and come back to the house with a bucket of water, several times a day.[2] Women typically do the laundry. On a clear, sunny day they fill big wooden tubs and start washing piles of dirty clothes, removing the crusty mud with a brush. They wash, rinse, wring out the water and hang them up before the sun disappears. Soon it will rain again and clothes must be slowly dried by hanging them above the stove.

Women are in charge of the preparation of meals, unless they are gone on a trip, and the man is left to carry out these tasks. They provide the food to be cooked, such as when they go to the beach to collect seashells. Whenever the tide is good, they go to the beach early in the morning, with a shovel and a basket. It is crucial to have *algo pa' echar a mi olla, para hacer mi comida* ("something to put in my pan, to make my meal"). The use of the possessive adjective underscores that this is "their" task—a proud statement of ownership and belonging. And every other day they make bread. Besides feeding their family, women also feed the livestock. Resourceful women know how to weave, knit, and make baskets. Blankets and baskets, besides being goods for the household, can also be sold or bartered.

Hard working, skilled, responsible men, and industrious, resourceful, and jocose women are considered good partners and desirable spouses. People with these characteristics have a better chance of being involved in a relationship and are appreciated and esteemed within the community. On the other hand, laziness, capriciousness, irresponsibility, and bitterness—as well as overindulgence in alcohol consumption—are character traits that are openly condemned (Image 7.2).

Women, men, and children work together in the field and in the gardens for every phase of the agricultural cycle: the preparation of the field before and after sowing, the sowing, and the harvest. They all take active part in the making of apple cider (*chicha*), as well as in the pig slaughtering and subsequent meat processing. They also work together intensely at the beach collecting seaweed in summertime (Morales and Calderón 2011). Collecting seaweed (*luga*) is a lucrative, seasonal activity that involves all family members including young children, keeping them busy starting at sunrise.

[2] This happened until mid-2015, when the local government provided running water to each island household.

Image 7.2 A couple in Apiao sows their potato field. (Photo credit: Giovanna Bacchiddu)

Most of the work described and listed above is done at regular times, within the same space, and interchangeably between women and men who consider their work complementary, like Harris' classic description of the Laymis' sexual division of labor (2000: 164–179). However, it is worth noting that Apiao residents generally do not regard occupations as *strictly* gendered: I have seen men kneading dough for bread or sweets, as well as knitting and sewing, and women transporting heavy pegs and uprooting roots from the ground. The emphasis is always on the household as a productive unit, to which each member of the family contributes actively. All work benefits the entire household, not just individuals.[3]

Going out

I now turn to what I call "social" situations, directing my attention to youth in particular. I will emphasize young people's attitudes toward sexuality and marriage and ideas about the status of being single.

Whenever islanders emerge from their houses, they are leaving their private realm to go outside, *salir pa'fuera*. The most visually apparent difference between someone involved in an activity within the house (or its immediate surroundings) and beyond the house is in their clothes. When working in and around the house, people generally wear work clothes (old trousers and wool sweaters for men and long aprons for women) and work shoes, usually rubber boots. Given the character-

[3] The only exception to this is the *luga* (seaweed) collecting: its profits generally benefit the individual collector.

istics of the terrain—extremely muddy for most of the year—and the rainy weather, clothes and boots quickly get dirty, and everyday activities lead to stains and patches. But whenever they go out, Apiao residents dress up in their best clothes and best shoes, kept for the relatively rare social occasions. People wash and carefully comb their hair, sport fashionable jeans, immaculate jackets, and shiny shoes.[4] Women bathe their children, dress them in new, elegant clothes and perfume them with cologne. Dressing appropriately for social occasions is considered an important social skill, a rule that children learn from an early age.

On Sunday evenings, religious *fiestas* and special occasions such as football (soccer) tournaments or dance parties, young people enjoy socializing with peers, away from their parents. Parents of teenage girls tend to be wary of their freedom and quite strict; girls must obtain permission before spending an evening out or going to a party. Reasons for this are seldom articulated, but clearly involve the sexual freedom young people enjoy at such gatherings. At the same time, older teenagers are thought to be of the right age for marriage and are expected to think about marrying soon.

Teenage girls who receive the attention of boys and young men decide whether to accept or reject the advances. Generally, the man takes the initiative, asking the woman to start a relationship, *pedir pololeo* in local Spanish. Although women would never ask *pololeo* of men, they flirt freely and show interest. Relationships usually start secretively and continue with the complicity of the friends and siblings of each party. Couples meet secretively for months, even years, before making the relationship public by establishing co-residence. They meet in the dark, spend time together, sometimes have sexual encounters, and go back home. They either lie to their parents to explain their absence or go out while everyone in the house is asleep. A young girl might have several relationships, one after the other, unless she is perceived as "difficult," i.e., not willing to accept courtship, joke, flirt, take on risks, and so on. A girl who continuously refuses courtship may be considered difficult and spoiled by her peers.

The shift from a hidden relationship to a publicly acknowledged, formal union between a man and a woman often depends on the man's initiative (Bacchiddu 2012a). It is the man who initiates a separate family by going to the woman's household and formally asking for her hand. This event, called *pedimiento* (literally, the "asking for [a wife]"), involves a great deal of formality. Although civil and religious marriage ceremonies may not be performed for months, the new couple is, effectively, considered married immediately after the *pedimiento*, when the woman leaves her family to live with her partner in his parents' house. There is little emphasis on marriage ceremonies, whereas there is a stress on the reciprocal obligations that tie individuals and families together.

[4] I was scolded a few times for wearing everyday clothes and rubber boots in public situations, which is considered counter-aesthetic and particularly inappropriate. Men who dance in religious occasions would excuse themselves if doing so with their boots.

What follows is the description of the relationship of a young couple, starting from the early stage of their hidden *pololeo* to the *pedimiento*, co-residence, and official marriage.

Anita and Pablo

Anita and Pablo had been seeing each other secretively for a few years. They were neighbors, growing up in two households only a few minutes' walk from each other.

Until the day she "was asked" by Pablo, Anita lived with her widowed mother, her grandmother, and her four unmarried siblings. Anita and her sisters used to do everything together: working in the fields and at home, visiting their married sisters or other relatives, and traveling to the nearby town to run errands. They were close and looked inseparable. In their mid to late twenties, they were remarkable for their discreteness, shyness, and their silent presence. I had the chance to closely observe the family interactions because during my fieldwork in 2005 I lived with Anita's family and was very close to Pablo's family.

Besides enjoying good neighborly relations, the two families were connected through the *compadrazco* system, which ties people in alliance and reciprocal obligations. Anita's younger brother was "given"[5] as a godchild to Pablo's father, making both sets of parents *compadres*. Pablo, moreover, was a good friend of Anita's younger brother, and together with his own brother they all played in the same football club every weekend. They often met to play cards and helped each other in various tasks. It was somehow easy for the couple to meet; they often relied on their younger brothers to pass on messages and cover up for them in case of a prolonged absence.

On the occasion of the launch of Pablo's new boat, he had formally requested[6] Anita's help to cook at his home for his friends who had helped him in the launch. Two of those friends were Anita's brothers. Anita accomplished her role as helper with respect and goodwill, mostly in silence. At the end of the lunch Pablo came into the kitchen with two bottles of wine: one for the table, and one for Anita. "This is for you!" he told her, handing her the bottle with a wide smile. Taken by surprise, she blushed, then accepted the bottle, looking away. Giving her a present in public was clearly a meaningful act, reflecting a deep attachment.

One evening I saw Pablo at a post-football charity event. He told me with a wide smile that he was going to marry soon: "On Friday I'm going to take her home! (*Me la llevo a mi casa.*) I came here to look for the *padrino*! It's going to be don Jorge."[7]

---

[5] To "give one's child" to a set of adults [to become their godchild] is the local way of saying that the child is godchild to the couple; the expression evokes a gift transaction and as such ties the giver (the child's parent) to the receiver (the godparent), starting a series of presumably lifelong obligations between the two parties.

[6] On the theme of formal requests, also see Bacchiddu 2010 and Bacchiddu 2012a, b.

[7] Don Jorge happens to be Pablo's father's brother. However, there is no prescriptive rule on choosing relatives as *padrinos*.

Pablo was referring to the *pedimiento*, the asking for the hand in marriage of a woman.

The *pedimiento*

On the *pedimiento day* I found Anita's family members busy with their usual chores, wearing work clothes and rubber boots. Her elder brother—the man of the house—had just interrupted his work: unshaven, he was wearing dirty clothes. *No pasa na'!* "Nothing happens!" he replied to my queries. The younger brother was busy playing football, I was told.

It seemed to me that all of them were doing their best to pretend nothing important was going to happen. However, the tension was palpable. Anita, her hair still wet, was nervously looking out the window. I asked her how she felt, and she said it was difficult to leave her house forever, that one never knows how things will turn out eventually.

When the guests arrived they were made to sit in the living room. Pablo's mother and the *madrina* were sitting together; the two men, father, and *padrino* were next to Pablo, who was extremely pale and stared constantly at the floor. Anita, also pale and silent, sat on the other side of the room. Pablo had brought a bottle of liquor and some glasses. He was supposed to serve the drink but was clearly too nervous, so he did it at random moments and nervously. Anita's mother entered the room in her working clothes, shook hands with the guests and sat down. Someone commented on the bad weather. Anita's mother replied politely but did not engage in the conversation, keeping her head low and holding her hands across her apron. Silence fell again. After several long pauses, the *padrino* started to talk in a serious tone, drawing out the words: "Well. We came for something, so now we will discuss the matter, because this is the purpose of our visit. *Señora*, we came after being asked by the young man here. The young man here asked us, as *padrinos*, to come and ask for Anita's hand, in the name of Pablo!"

I lifted my gaze and realized everyone was humbly looking down, with the young couple especially embarrassed, and pale. Even don Jorge, despite his confident manner of speaking, had a very humble and embarrassed expression in his face.

Anita's mother replied curtly and bluntly; she was clearly reluctant to let her daughter go. Don Jorge spoke again, saying that having accomplished their duty, they could go: "*A eso vinimos, y ahora nos iremos andando*," and they all stood up, ready to leave. Anita's mother asked them to wait for a small dinner that was being prepared in the kitchen. The table was quickly set and the three couples sat to eat behind closed doors: Pablo and Anita, their godparents, and Pablo's parents. No one else sat at the table with them, and no one stayed in the living room. Anita's family were having the same meal but in a different room, the kitchen. The two rooms were separated by a wall, and a door that was kept shut. While eating in silence, we could hear the guests next door chatting and occasionally laughing.

When they finished their meal, the guests stood up, gave their thanks, and left hastily, using the main entrance of the house reserved for special guests, not the kitchen entrance. Anita joined them, carrying a small suitcase. That moment marked an important transition: she was leaving her home to live with her husband in the house of her in-laws. From that moment on, she became an independent, autonomous person, detached from her own family and household. From the moment her mother agreed to "give" her, Anita became part of the guests' group: being served, rather than serving, unlike her unmarried sisters. Yet, nothing seemed to mark this event. Anita's sisters left straightaway to go and continue with their work, the elder brother sat with his mother and grandmother behind the stove, as if it were any other evening.

In the groom's household, where I later joined them, the kitchen table was set and beef was roasting in the oven. We all sat at the table for our second dinner. The atmosphere was relaxed and the TV was on. We all ate in silence while watching a soap opera on the national channel, a favorite of every household. We chatted about several topics, none related to the marriage. Pablo, radiant, served wine to everyone, filling glasses as soon as they were emptied. Abundance of wine is usually a sign of a special celebration, as was the grilled beef, served instead of the more common stew. Clearly, the bride's family had prepared the dinner hastily to attend to the guests, rather than to celebrate the marriage of a family member. Far from being joyous, they seemed to project a sense of loss.[8] Even the groom's family was demure, with few outward signs of celebration.

The reticent atmosphere surrounding the event was remarkable. Despite my familiarity with both families, and my presence in their households during the preceding weeks, the subject of the impending wedding was never broached. The groom's parents seemed to deliberately avoid the topic; indeed, it was downplayed by both parties: no special clothes were worn, no special food was prepared, and no time was devoted to discussing wedding preparations. Activities continued as normal. Everyone watched the TV soap during dinner, as if to underscore that this was an ordinary day.

The tradition of the *pedimiento* is a good example of what Van Gennep calls "social disturbance," a "disturbance of equilibrium" (1960:139), in what seemed to me, at the time, more an intrusion and an interference of the daily routine than a transition celebration.

---

[8] Confront with the following passage described by Van Gennep in his seminal book on rites of passage: "To marry is (…) to pass from one family to another (….). An individual separation from these groups weakens them but strengthens those he joins. The weakening is at once numerical (and therefore a reduction in force), economic, and emotional" (Van Gennep 1960:124).

A detailed account of Mapuche weddings among "civilized indigenous people" as opposed to traditional weddings through kidnapping is described by Noggler (1982: 14ff). What appears significantly similar is the sequence of events, and the formality of the speeches between the parties involved – notably, the mediator (the part that belongs to the *padrino* in Apiao) and the father of the bride. Also, confront Faron (1961: 156ff), where marriage by elopement among the Mapuches is described.

Getting Married: Carmen's Story

Months before Anita's *pedimiento*, I spoke with Carmen, Pablo's mother, about couples and *pololeo*, and she told me her story. She said that in the past, couples had to hide, because if the parents found out they were dating, the young girl would be beaten. No girl was allowed to go out by herself, so she needed the company of friends to keep the affair secret. Carmen met her future husband in the church square. They secretly dated for about a year, meeting sporadically when there was an event near the church, *cuando había movimiento*. One evening, he came to take her home with him. He had warned her that he would do this, but she had not taken him seriously.

She had spent the day working in the fields, and when she got home she found the kitchen full of guests: *Una tremenda cuadrilla e' gente!* It was her lover with her future mother-in-law, and the *padrinos*. Her parents were very upset with her—*Me enojaron harto*—but they could not refuse her to them. "You cannot deny a daughter when she is asked for with *padrinos* and everything!"

Why were her parents annoyed? Was it because they realized she had been carrying on a secret relationship? Or was it because of the prospect of losing her labor (see Van Gennep 1960: 124)? When a woman is asked for by a man, or rather by a *padrino* on behalf of a man, her family faces a certain pressure to acquiesce to the request. A refusal would be seen as disrespectful toward the *padrino*, the prospective husband, and the husband's parents. The night Carmen was asked for by her future husband and his *cuadrilla*, she went home with them straightaway, *Con lo puesto,* literally meaning "with the clothes she was wearing."[9] I asked if she had carried a bag of clothes, and she said that she did not have time for that. She had to go *con lo puesto*, because, "When they come to take you, you must go." In the following days she went back to look for clothes. Then I asked if she was given a room for herself in the groom's house. "Of course we slept together!" she said, "*Al tiro*!" "Right away!"

She didn't know her husband's family well before getting married, *De día y de tarde no más.* They would greet her politely, from day to day, and say little else. During the co-residence, her in-laws were kind to her. After a week, though, she began to miss her family. A great distance separated her from her original household, and at that time there were no paths on the island. To reach near and distant places one had to cross private fields and thick vegetation. Today, a dirt road bisects the island and facilitates mobility. But when Carmen got married, about 40 years ago, it was difficult to visit her parents frequently. Eventually, she got used to her new life.

Another episode shows how the institution of marriage in Apiao is downplayed and minimized. Antonia, a woman in her early fifties, once invited me to join her family for a special dinner. They were celebrating her husband and her elder child's

---

[9] People use this expression to describe something that happens unexpectedly, situations in which people cannot possibly change their clothes, a matter of concern when people go out for various events.

santorál, name-day. They liked to remember the day with a special meal—good food and abundant alcohol. "I always remember this day," she told me, "and today marks 23 years since I entered this place for the first time." Yet, the celebration was for the name-day, not for the *pedimiento* anniversary. During dinner the family mentioned several other name-day celebrations of the past, and the couple's child told me how sad he had been once on this day, forced to eat beans on his own, while doing his military service. No mention was ever made of how and when the couple started their life together.

The New Couple

After that first night spent with her future husband at her in-laws' house, Anita became an integral part of Pablo's family. She helped her mother-in-law with the daily routine; she prepared meals, made bread, and attended visits, like a proper house owner, *una dueña de casa*.

Typically the postmarital residence pattern in Apiao is patrilocal, as in the case described, with the woman going to live with her in-laws. But there are exceptions to this. In some cases, the husband moves to the wife's household, either because the man cannot provide good accommodation and land for cultivation, or because the woman owns fields and there is no one else to tend them.

The postmarital residence is generally provisional: the new couple will live with the man's parents until their new house is built. After a period ranging from months to years, the man typically builds a new house for his nuclear family, often on his parents' property. In the period preceding the move, the couple will try to save money and accumulate household goods to enable them to live independently. They also often have one or more children by the time they move in to a new home.

The construction of a new house is strictly associated with a marriage. If a bachelor builds the house while his parents are alive and still living in their home, it generally means he is planning to get married soon. If he doesn't marry, he may be ridiculed by the community, and the new house may even be abandoned, since few bachelors would be willing to live on their own.

Casarse por las dos leyes: Getting Officially Married

Usually months or years separate the *pedimiento* and co-residence from the civil and religious wedding ceremonies. Sometimes couples do not officialize their union until they are asked to be *padrinos* (either for a wedding, baptism, communion, or confirmation). To obtain permission from their Catholic parish, they must be officially married by both the state and church. This is one way the Catholic Church has been able to perpetuate the institution of marriage—both civil and religious—in rural Chile.

This was precisely what forced Pablo's parents to marry in haste. They were asked by neighbors to be *padrinos* for a confirmation ceremony, but they had not been married by the church. The bishop of the diocese happened to be visiting the island at the time and, since the groom was a catechist, he agreed to marry them in a private ceremony the night before the confirmation ceremony. The couple's recollection of the events stressed the fact that they *had* to get married in order to honor the pact they had with their godchild's parents. It was, from their perspective, a nuisance, a potential problem solved by the generosity of the bishop, who granted a last-minute ceremony to render the union official. Indeed, I was told that the wedding *padrinos* are the ones held responsible for reminding their godchildren to validate their union with both the state and church.

Residents of Apiao use the expression *por las dos leyes* to describe a marriage that has been officialized through church and state ceremonies. Indeed, they will often state the words emphatically to stress the official status of the marriage. The two ceremonies are generally held on the same day. The date must coincide with the presence of a priest on the island. Although priests might travel to Apiao once or twice a year, the local parish usually sends a priest on or around the occasion of the island's patron saint celebration, on the 11th of February. Several religious ceremonies may be performed on this day, including weddings, baptisms, communions, and, if a bishop is present, confirmations. The couples willing to get officially married travel early in the morning to the closest civil registry along with their *padrinos*. The registry is on a nearby island—Quichao—a few hours away. After the civil ceremony, they return to Apiao and attend the mass, where they are married in a public wedding ceremony.

While few people accompany couples and *padrinos* to the civil registry, the religious ceremony is performed in the crowded wooden church of the island. Couples, children, their parents, and *padrinos* all gather in front of the altar, while fellow islanders fill the pews. The event is intermixed with other ceremonies; these often stress the strengthening of ties and alliances between families and individuals more than the change of status of the couple. Moreover, all of the ceremonies are viewed as an extension of the island's annual religious *fiesta*.

Wedding Party at Home: Offering Dinner to *Padrinos* and Guests

Because the distinction of public wedding celebrations is somehow blurred in the atmosphere of the patron saint festivities, most couples organize a special dinner for their families and *padrinos*. This can be an ambitious event involving a great deal of expense. Consequently, some couples limit their offerings of food to the godfathers, to whom they are indebted, since the *padrinos* accepted the groom's request. But whether the celebration is lavish or simple, it is seen as a necessary reciprocal act: the *padrinos* must be thanked for having agreed to fulfill an important social role for the young couple.

Family and friends come to these meals expecting to relax and enjoy food and drink. *Todos mis hijos se casaron como corresponde* ("All my children got married the way it should be done"), a woman told me, reminiscing about the wedding parties she had organized. She listed the dishes that were prepared at her children's wedding dinners: appetizers, chicken soup, roast beef, potatoes and salad, and a three-layered cake, "Enough for everybody to eat," and abundant wine, liquor, and soft drinks. This variety contrasts with everyday meals that, while plentiful, consist of one main dish, usually a stew, served with boiled potatoes. Dessert is rarely served, and alcohol is purchased for special occasions only.

Food is not only offered but also given away. Forty kilos of wheat are turned into bread to be distributed to all guests. This offering of bread usually marks an exchange which is both material and ritual: bread is given both as a form of payment for work being performed, and as a sign of gratitude to participants in funerals and other religious events held in private households. Bread is also given to participants at nonreligious festivities,[10] as in this case.

Anita and Pablo organized, with help from their families, what was widely viewed as "a proper" dinner party. Guests were expected to bring a present, to be given to the spouses discreetly, out of sight of others. The newlyweds were called outside the house by different guests for this purpose. In general, presents on these occasions are neither shown nor discussed in public.

While all guests receive bread as a token of appreciation for their presence and their gifts, a more conspicuous exchange occurs between godparents and their marriage godchildren (*aijados de casamiento*). The *padrinos* are expected to bring a gift for the couple, and the couple is expected to give the *padrinos* meat and alcohol in return, generically referred to as *la atención*. Pablo and Anita received from their *madrina* a boxful of kitchenware: a tea set, dishes, glasses, cutlery, and pans. They also received a female lamb from the *padrinos*.

"I remember when I got married," one woman told me. "My *padrinos* gave me nothing! Not even an old kettle, or a pan…I had to start from scratch, since I was given nothing. Just a sheep."

Objects bought in town and industrially produced are valued more than presents such as sheep or lambs that, while certainly appreciated, are readily found on nearly every rural household. The idea behind the *madrina*'s gift was to provide them with the essentials to use once they moved into their own house. But no matter the present, the couple was expected to pay their debt to the godfathers with the *atención para los padrinos*. This consisted of half a pig, a bottle of spirits, and several liters of wine to the *padrino*; a chicken and a large bottle of soft drink to the *madrina*, along with several one-kilo loaves of bread cooked in a special oven. "I went there

[10] At the end of my fieldwork my host family organized a goodbye dinner party for me. The guests were attended with abundant food, alcohol, and, at the moment of leaving, each guest was given one loaf of bread and some roasted meat to take away. This, it was explained, was done to thank them for their presence at the dinner party, and, as my hosts put it, *pa'que se vayan conformes*, "for them to leave satisfied."

with a heavy load, and I came back home with an even heavier load!" the *madrina* later told me. She was given so much meat, that she ended up distributing it to her neighbors.

The dinner was followed by a dance session with live music. A room was cleared to provide space for dancing that continued until the early morning. Throughout the night, wine was served to all guests and to the musicians.

Most of the gifts—excluding the *padrinos*' store-bought gift to their godchildren—and most of the food that was consumed—except the soft drinks—were local products from the couple's household. This reflected the importance of owning land, livestock, and crops. Only with recourse to such property could gifts be offered, relationships initiated, and alliances perpetuated by the new family.

Relation between *Padrinos* and *Aijados*

Marriage and its celebrations can bind together as many as four couples: the newlyweds, their parents, and the *padrinos*. After the wedding, the relationship between the *padrinos* and their *aijados* is one of *obligación* (obligation). This is also the case with the parents of the bride and groom, who become *compadres* to each other and to the *padrinos*. This relationship stipulates that they address their new *compadres* not by their name but by the titles *compadre* and *comadre.* When talking to others they are also expected to use the titles (always preceded by the possessive adjective "my," or *mi*). The married couple is expected to do the same when addressing or referring to their godparents, using the words *mi padrino* and *mi madrina*.

Being in *obligació*n to someone implies paying special attention to them and acknowledging a long-standing commitment to help them. In the case of *aijados* and *padrinos*, the *aijados* are expected to host them properly during the marriage ceremonies. During the annual pig slaughtering they are expected to give the *padrinos* meat, bread, and processed pig products, called *lloco*. The *padrinos* are expected to reciprocate the gift, once they slaughter their own pig.

The Meaning of Marriage: Productivity and Independence

After her *pedimiento*, Carmen lived with her husband in his parents' household for 5 years. Later, the couple moved to the other side of the island and built a house on land belonging to Carmen's parents. They hoped to harvest more crops on the new land. For 5 years they had farmed on plots near the house of Carmen's in-laws. Now they had more space for agriculture. "This is why we came here," she explained. "For the land."

As expected in a quasi-subsistence economy, productivity in Apiao is critically important and individuals become independent through productivity. A new couple, in other words, is first and foremost a new productive unit. Upon initiating co-residence with in-laws, a couple sets aside a portion of land that will become theirs once they have the resources to build their own house. Despite sharing a home, meals, time, space, and work with relatives, the life of a couple is a continuous affirmation of their independence. Indeed, individuals in Apiao are taught to be independent from an early age. Children are expected to help with work in the home and the fields. Upon marrying, individuals have the opportunity to assert their independence and double their productivity.

Prior to marriage, individuals often work on behalf of their parents. After marriage, they speak for themselves and assume responsibility for their own productive unit. It is in separating themselves from their immediate family and forming a separate household that they can claim the status of independent nucleus. In this regard, marriage enables individuals to become active social actors. Established couples can take the role of hosts, formally invite individuals to take part in work parties, thereby initiating reciprocal relationships. The first of these is their relationship with their godparents. Others soon follow: work relationships with particular neighbors, kin obligations, other *compadrazcos*.

When a couple has a child, they choose another couple to be the child's baptism godparents, establishing yet another important relationship of *obligación*. These relationships are reaffirmed during religious celebrations such as Holy Communion, Confirmation, and so on. Married couples, in this way, build and strengthen exchange networks that provide not prestige and social influence, but rather a continuity of social life between equals.

After living together for 9 months, Pablo and Anita hosted their first pig slaughtering, called *carneo de chancho*. This important event, taking place annually in nearly every household in Apiao, mobilizes neighbors and kin who work together intensely for a day. A few days before the event the hosts visit neighbors to formally invite them to the *carneo*. I was impressed to see both Anita and Pablo ride horseback through heavy rain to the house where I lived to invite all of us to the event. It was one of their first social assertions of independence. The year before, Pablo had come to extend an invitation on behalf of his parents; this time he was acting for himself and his wife.

On the day of the *carneo*, Pablo insisted on killing the pig himself by being the one to slide the knife into the pig's heart. He did this under his father's supervision and with his friends watching and giggling. Given that it was his first time, he did not guess the right spot and the pig took a long time to die. Pablo, embarrassed and ill at ease with his new role as butcher—carried out somewhat ineptly as family head—joked with his friends and family as they sweetly teased him over his performance. "I only feel bones!" he said laughing, and the *carneo* site resounded with affectionate laughter.

In the days following the *carneo*, Pablo and Anita traveled to several households (notably, those of their *padrinos* and *compadres*,[11] and to Anita's mother and sisters) to offer a gift that consisted of large cuts of pork and various processed pork prod-

[11] The baptism godparents of their child.

ucts. Called *lloco* or simply *plato 'e comida* (food plate), this gift underscores the donor's independence and productivity, and establishes or strengthens the reciprocal relationship between the two parties.

A few days later I witnessed a formal expression of Anita's independence. I had accompanied Anita's mother and brothers to their own *carneo*, and I was present when Anita's mother went to her daughter's new residence with a heavy load of *llocos*. Anita welcomed her in the way every guest is welcomed, by preparing *mate* and serving it to the guest with freshly sliced bread, served on a plate on top of a stool. After having eaten the bread and sipped the *mate*, the woman handed Anita a large wheat sack full of pork, saying, "Anita, this is for you." Anita placed herself in front of her seated mother and asked for her hand, addressing her with the formal pronoun *usted*.

"Show me your hand," she said. "Thank you so much for taking the trouble [to give me this gift], may God accompany you and give you [good] health."[12]

These words, spoken while holding the giver's hand, are uttered every time a *lloco* is given. Indeed, the same words are uttered, in the same manner, every time a gift or an offer is given and received. After Anita had spoken, Pablo thanked his mother-in-law in exactly the same way.

I was initially puzzled by the formality between mother and daughter. They had lived together and shared everything for 28 years. I asked several people about this and was given the same explanation: "Anita is part of an independent family. She does not belong to her mother's family anymore. Therefore, it is as if she were any other person, and she must thank in that form."

Marriage in Apiao, and in much of rural Chiloé, enables individuals to become *familia aparte, familia independiente*—separate, independent family—and to express and carry out all their productive potential. Anita's formality was a statement of her independence, of her autonomy gained through marriage. Indeed, the entire *carneo* event at the new couple's home was a statement of independence: the preparation that required them both to personally invite individuals and formally ask for help; the actual pig slaughtering that required an effort by Pablo to execute a task that he was ill suited to do; and the final act of distributing *llocos* and accepting them as an independent, autonomous family (Image 7.3).

Once individuals leave their family home, they must assume the responsibility that comes with an independent household. They are bound to continue the cycle of productivity. A man and a woman form a complementary unit, working jointly toward productivity. Sometimes one of the two individuals leaves much of the workload to the other person, who is then left with a heavy burden. This problem can affect both men and women, and Apiao residents often say that nothing can be done about it. Life goes on, they declare. Priority, nevertheless, is given to autonomy, an immediate consequence of productivity. Once individuals become *familia aparte*—separate family—their independence cannot be undone, and with or without the

[12] *A ver su mano. Muchísimas gracias 'onde se molestaron, que Dios la acompañe y les de la salud'.*

Image 7.3 A family plowing their field with a *yunta* (yoke). (Photo credit: Giovanna Bacchiddu)

help of their spouse, they have to cope with the daily workload necessary for the successful running of the household. "It's like with the yoke," a woman once told me. "The oxen don't all pull with the same effort!"

The image of the *yunta* (yoke) is indeed an appropriate one for Apiao. As the wooden crosspiece that is fastened over the heads of oxen, the *yunta* is one of the most important work tools for islanders. It is used to help plow potato fields, pull carts, and transport heavy objects. The two beasts of burden, horns fastened together, work in unison. If one of the animals is sick or injured, the *yunta* cannot be used, and the oxen are useless.[13]

Conclusions

Marriages in Apiao are, more than anything else, about the establishment of a new productive unit. The ceremonies surrounding marriage emphasize the shift that a couple undergoes in moving from one productive unit (the parents' households) to another (the couple's new household). The married status is accompanied by an assertion of independence by the couple, symbolized by the couple's new agricultural productivity.

In Apiao individuals, not groups, are the basic social unit, and all transactions happen between individuals, or sets of individuals, such as married couples. A mar-

[13] See Olivia Harris (2000: 33): for the Laymi, the bulls joined in a yoke are an expression of complementarity: "their paired duality under the yoke makes of bulls a primary expression of the integral bond between humans and earth."

riage entails negotiation between individuals: a husband and a wife; a woman and her godmother; a man and his godfather; a person and his or her in-laws. Individuals are valued for their ability to turn their work into productivity, and their household into an independent, autonomous unit. To achieve that goal, the ideal that is pursued is the unity and complementarity of the married couple, which is symbolically likened to a yoke of oxen—the main tool used in Apiao's agricultural activities. In this regard, a pair of individuals is seen to become a single, productive force.

Dedication This chapter is dedicated to the memory of *abuela* Elisa Millalonco Barria, beloved Apiao grandmother and friend.

References

Bacchiddu, G. 2010. Getting Tamed To Silent Rules: Experiencing 'the Other' in Apiao, Southern Chile. In *Mutuality and Empathy: Self and Other in the Ethnographic Encounter*, ed. A.S. Grønseth and D.L. Davis, 21–34. Wantage: Sean Kingston Publishing.

———. 2011. Holding the Saint in One's Arms. Miracles and Exchange in Apiao, Southern Chile. In *Encounters of Body and Soul in Contemporary Religious Practices. Anthropological Reflections*, ed. A. Fedele and R.L. Blanes, 23–42. New York and Oxford: Berghahn.

———. 2012a. Doing Things Properly': Religious Aspects in Everyday Sociality in Apiao, Chiloé. In *Ordinary Lives and Grand Schemes: An Anthropology of Everyday Religion*, ed. S. Skielke and L. Debevec, 66–81. New York and Oxford: Berghahn.

———. 2012b. Mães Relutantes, Avós Maternais e Pais Esquecidos: o Fazer e Desfazer das Relações de Parentesco em Apiao, Chiloé. *Revista Tellus* 23: 35–58 (jul./dez. 2012).

Faron, L.C. 1961. *Mapuche Social Structure*. Urbana: The University of Illinois Press.

Gow, P. 1995. Land, People and Paper in Western Amazonia. In *The Anthropology of Landscape: Perspectives on Place and Space*, ed. E. Hirsch and M. O'Hanlon, 43–62. Oxford: Clarendon Press.

Harris, O. 2000. *To Make the Earth Bear Fruit: Essays on Fertility, Work and Gender in Highland Bolivia*. London: Institute of Latin American Studies.

Morales, C., and M. Calderón. 2011. *De Booms y Fiebres Marinas: Breve Historia Económica de Isla Apiao y el Mercado de Algas*. Santiago de Chile: CNCA.

Noggler, A. 1982. *Cuatrocientos Años de Misión Entre los Auracanos*. Temuco: Imprenta y Editorial San Francisco.

Van Gennep, A. 1960. *Rites of Passage*. Chicago: University of Chicago Press.

Name Index

A. Daughters, A. Pitchon (eds.), *Chiloé*, Ethnobiology,
https://doi.org/10.1007/978-3-319-91983-6

Index of Places

A. Daughters, A. Pitchon (eds.), *Chiloé*, Ethnobiology,
https://doi.org/10.1007/978-3-319-91983-6

T

V

Subject Index

A. Daughters, A. Pitchon (eds.), *Chiloé*, Ethnobiology,
https://doi.org/10.1007/978-3-319-91983-6

FSC
www.fsc.org
MIX
Papier | Fördert
gute Waldnutzung
FSC® C083411

Kolibri 360 GmbH
Ferdinand-Jühlke-Straße 7
99095 Erfurt, Deutschland
produktsicherheit@kolibri360.de